AEROSOLS

SYNTHESIS, OPTICAL PROPERTIES AND ENVIRONMENTAL IMPLICATIONS

ENVIRONMENTAL HEALTH - PHYSICAL, CHEMICAL AND BIOLOGICAL FACTORS

Additional books in this series can be found on Nova's website
under the Series tab.

Additional e-books in this series can be found on Nova's website
under the e-book tab.

ENVIRONMENTAL HEALTH - PHYSICAL, CHEMICAL AND BIOLOGICAL FACTORS

AEROSOLS

SYNTHESIS, OPTICAL PROPERTIES AND ENVIRONMENTAL IMPLICATIONS

ADAM YANICK PEARSON
EDITOR

Novinka

New York

Library of Congress Cataloging-in-Publication Data

ISBN: 978-1-63117-512-1

Library of Congress Control Number: 2014934031

Published by Nova Science Publishers, Inc. † New York

CONTENTS

PREFACE

Nanostructured materials and coatings have gained great importance due to their microstructural properties. Their applications have increased in different technological areas such as, in environmental pollution control, photo catalysis, optics, solid oxide fuel cells, electronic and optoelectronic devices, mechanical protection, catalysis, and in biomedical. Furthermore, advances in aerosol processing in recent years have increased the variety of produced nanostructured materials, including metals, oxides, ceramics, and composites, in different appearances such as nanoparticles, rods, belts, fibers, tubes, needles, and films. This book discusses the synthesis, properties and implications that aerosols have on the environment.

Chapter 1 - In assessing the effect of aerosols on climate change, it is not only the total aerosol loading, but also their absorption properties that determine the magnitude and the sign of aerosol forcing. Absorbing aerosols, such as black carbon, contribute to global warming by directly absorbing solar radiation and indirectly interacting with the hydrological cycle. The absorption property of aerosols is usually characterized by the absorption optical depth (ABS) and single scattering albedo (SSA). However, the temporal variability of these two parameters, as well as the implied aerosol compositional changes, has not been well studied, primarily due to the incapability of state-of-the-art satellite sensors to retrieve these properties. In this study, the author uses ground-based observations from 70 carefully selected stations within the Aerosol Robotic Network (AERONET), the currently most extensive and highly accurate surface aerosol monitoring network, to investigate the temporal trends in aerosol absorption and composition during the past decade. A weighted linear regression is applied to account for the error and outliers in the measurements, especially those in Level 1.5 data. My results indicate that during the past ten years, global ABS measured at 675 nm is dominated by a decreasing trend of approximately -0.003/decade, associated with an increase in SSA at ~0.03/decade, while no significant trend in aerosol extinction optical depth is found. The trends are most significant for the US, South America, Europe and East Asia, but not significant for Africa. By classifying the stations into five representative aerosol types, statistically significant decreasing ABS trends are found for all four aerosol types except for dust, which suggests decrease in black carbon is likely responsible for the observed trend. This hypothesis is further supported by an EOF analysis of the Joint Frequency Distribution (JFD) of Angstrom Exponent (AE) and Absorption Angstrom Exponent (AAE), which is an effective characterization of aerosol composition. The results clearly indicate a decreasing trend in the mode corresponding to black carbon. Furthermore, by sorting the data according to the black carbon AE/AAE cluster to minimize the influence of dust, the Africa stations also show significant positive SSA/negative ABS trends. The trends presented here are consistent with previous model simulations and projections for some regions, especially US and Europe. However, more research is likely required to find the cause of this black carbon reduction for all regions.

Chapter 2 - The activities of the cosmogenic nuclide beryllium-7 (^{7}Be) in near-surface atmospheric aerosols simultaneously collected at five cities in the East Asian monsoon region during four seasons were measured using a high-volume air sampler and a high-resolution gamma-ray spectrometer. The latitudinal distribution of the annual average ^{7}Be concentrations in aerosols follows a normal distribution pattern, with the maximum (8.31 ± 0.49 mBq m^{-3}) occurring at ~ 40°N, which was significantly higher than values reported for other cities in the East Asian monsoon region and in the world during the same period. The simultaneous observation of ^{7}Be at different latitude sampling sites in the East Asian monsoon region also displayed approximately a normal distribution pattern, but with the maximum at 30°N in spring and autumn and at 40°N in summer and winter. At a particular time, the latitudinal distribution pattern can instantaneously move southward or northward depending on season. Higher mean ^{7}Be concentrations in spring and autumn could be attributed to the transition of atmospheric circulation patterns. On the other hand, near-surface atmospheric ^{7}Be concentrations were lower in winter and summer with northward shift of the peak position. Atmospheric circulation may also play an important role in the atmospheric long-range transport (LRT) of persistent organic pollutants (POPs), but the extent to which it affects the latitudinal distributions of POPs in atmosphere and surface soils remains uncertain. Using ^{7}Be as a reference, simultaneous observation of ^{7}Be and typical POP compounds at different latitude in the East Asian monsoon climate system (EAMCS) showed similar latitudinal distribution patterns, suggesting that atmospheric circulation can affect the latitudinal distributions of POPs in the atmosphere.

Chapter 3 - Nanostructured materials and coatings have gained great importance due to their microstructural properties and their use has increased in different technological areas such as: environmental pollution control, photocatalysis, optics, solid oxide fuel cells, electronic and optoelectronic devices, mechanical protection, catalysis, biomedical applications, etc. Furthermore, advances in aerosol processing in recent years have increased the variety of produced nanostructured materials, including metals, oxides, ceramics, composites, and fullerenes; in different appearances such as nano-particles, rods, belts, fibers, tubes, needles, films, etc. This chapter presents the synthesis of nanostructured oxides in the form of nanoparticles, and multilayered or composite thin films by aerosol assisted chemical vapor deposition technique. It is well established that atomic arrangement and the microstructure of materials determine their properties. Interestingly, similar morphologies cut across all material classes and the different methods used to synthesize them; thus, controlling composition, grain size, morphology, and crystallinity are primary concerns of the synthesis and processing of materials.

In consequence, a microstructural characterization of nanostructured materials and coatings is crucial in the optimization of processing in order to obtain a desired property or behavior. Therefore, a detailed microstructural characterization of these materials was performed by electron microscopy and x-ray diffraction. In the case of films and multilayers, surface micro-structure was analyzed by field emission scanning electron microscopy (FESEM); also, cross-sectional microstructure were also analyzed by FESEM and high resolution transmission electron microscopy (HRTEM). Cross sectional samples for HRTEM analysis were prepared by focused ion-beam technique. Elemental composition of the materials was determined by energy dispersive x-ray spectroscopy with corresponding systems attached to electron microscopes. Crystalline structure was analyzed by grazing incidence x-ray diffraction.

The optical characterization of the thin films was attained by UV-VIS-NIR spectroscopy; optical constants and thickness were also determined. The optical properties were discussed in relation to the microstructural characteristics of the samples. Some of these materials were utilized for environmental pollution control applications. Films were employed for liquid phase photocatalysis of model pollutants; and nanostructured powders were used for the removal of As from aqueous solutions.

In: Aerosols
Editor: Adam Yanick Pearson

ISBN: 978-1-63117-512-1
© 2014 Nova Science Publishers, Inc.

Chapter 1

RECENT TRENDS IN AEROSOL ABSORPTION AND COMPOSITION DERIVED FROM AERONET MEASUREMENTS

Jing Li[*]
NASA Goddard Institute for Space Studies, New York, NY, US

ABSTRACT

In assessing the effect of aerosols on climate change, it is not only the total aerosol loading, but also their absorption properties that determine the magnitude and the sign of aerosol forcing. Absorbing aerosols, such as black carbon, contribute to global warming by directly absorbing solar radiation and indirectly interacting with the hydrological cycle. The absorption property of aerosols is usually characterized by the absorption optical depth (ABS) and single scattering albedo (SSA). However, the temporal variability of these two parameters, as well as the implied aerosol compositional changes, has not been well studied, primarily due to the incapability of state-of-the-art satellite sensors to retrieve these properties. In this study, I use ground-based observations from 70 carefully selected stations within the Aerosol Robotic Network (AERONET), the currently most extensive and highly accurate surface aerosol monitoring network, to investigate the temporal trends in aerosol absorption and composition during the past decade. A weighted linear regression is applied to account for the error and outliers in the

---

[*] Email: Jing.Li@nasa.gov.

measurements, especially those in Level 1.5 data. My results indicate that during the past ten years, global ABS measured at 675 nm is dominated by a decreasing trend of approximately -0.003/decade, associated with an increase in SSA at ~0.03/decade, while no significant trend in aerosol extinction optical depth is found. The trends are most significant for the US, South America, Europe and East Asia, but not significant for Africa. By classifying the stations into five representative aerosol types, statistically significant decreasing ABS trends are found for all four aerosol types except for dust, which suggests decrease in black carbon is likely responsible for the observed trend. This hypothesis is further supported by an EOF analysis of the Joint Frequency Distribution (JFD) of Angstrom Exponent (AE) and Absorption Angstrom Exponent (AAE), which is an effective characterization of aerosol composition. The results clearly indicate a decreasing trend in the mode corresponding to black carbon. Further more, by sorting the data according to the black carbon AE/AAE cluster to minimize the influence of dust, the Africa stations also show significant positive SSA/negative ABS trends. The trends presented here are consistent with previous model simulations and projections for some regions, especially US and Europe. However, more research is likely required to find the cause of this black carbon reduction for all regions.

1. INTRODUCTION

Aerosols remain the largest source of uncertainty in anthropogenic forcing of climate change (IPCC, 2007). A major complication in aerosol direct radiative forcing arises from the opposite effect of scattering and absorbing aerosols, and their relative fraction. Scattering aerosols, such as sulfates and nitrates, cool the atmosphere and surface and thus counteract the global warming induced by greenhouse gases. On the other hand, absorbing aerosols including black carbon and mineral dust, heat the atmosphere and contribute to global warming. Whether changes in atmospheric aerosol loading will induce a heating or cooling effect on the climate heavily depends on the fraction of scattering and absorbing aerosols, quantified as the aerosol single scattering albedo (SSA), which is defined as the ratio of scattering optical depth to the total extinction optical depth. Changes in aerosol SSA impact heavily on both the direct effect and aerosol-cloud interaction. It is believed that there is a critical value of the SSA above which aerosols produce a negative forcing (cooling), and below which there is positive forcing (warming). Hansen et al. (1997) concluded from GCM experiments that aerosol mixture with SSA up to

0.9 would lead to global warming, and that the anthropogenic aerosol feedback on the global mean surface temperature is likely to be positive. Ramanathan et al., (2001) indicated that for SSA<0.95, aerosol net forcing can change from negative to largely positive depending on aerosol height, surface albedo and cloud conditions.

Moreover, as current satellite sensors lack the capability that is needed to measurement aerosol absorption optical depth (ABS) or SSA, these properties must be prescribed in aerosol models used in the retrieval algorithm. So there is no ability to tell from satellite data whether aerosol SSA remains constant, or if the aerosol SSA is in fact changing. Obviously, a changing SSA would bias the inference of any trends in global or regional aerosol loading as defined from satellite retrievals.

As a result, knowledge of temporal trends in aerosol absorption and SSA becomes essential to quantify the aerosol effect in global warming and to accurately estimate aerosol trends from satellite sensors. However, the study on SSA trends is still limited primarily due to the difficulty in retrieving this quantity from space. Currently, our best available information on aerosol SSA comes from the network of ground-based aerosol measurements using sun photometry (Holben et al., 1998). The Aerosol Robotic Network (AERONET), started mostly after 1998, provides ABS and SSA data (Dubovik and King, 2000) at four wavelengths (441, 675, 870 and 1020 nm) for over 400 stations globally, which appears to be suitable for performing regional scale SSA trend analysis.

In this study, I examine temporal trends of aerosol ABS and SSA at 70 carefully selected stations over the ten-year period from 2003 to 2012. Linear trends are estimated using a weighted regression to minimize the errors resulted from sampling and outliers. Furthermore, to infer aerosol compositional changes, we make use of the spectral dependence of ABS and SSA. Recent studies have shown that it is primarily the fraction of black carbon, organic matter, and mineral dust in atmospheric aerosols that determines the wavelength dependence of absorption, and thus the spectral signature of the single scattering albedo (Russell et al., 2010). Giles et al. (2012) indicated that aerosol types could be characterized by the Angstrom Exponent (AE) –Absorption Angstrom Exponent (AAE) cluster, where the AAE is the spectral dependence of ABS. An Empirical Orthogonal Function (EOF) analysis is performed on the AE-AAE joint density distribution to extract the cluster structure and temporal variability of aerosol composition.

The chapter is organized as follows: section 2 describes the data selection criteria and analysis method. Section 3 presents trend analysis for each individual station, and averaged by different regions and aerosol types. The discussion on the distribution of AERONET data and the rationale to use Level 1.5 data is given section 4. And section 5 summarizes the study.

2. DATA AND METHOD

Aerosol single scattering albedo, absorption optical depth and extinction optical depth (AOD) at 440, 675, 870 and 1020 nm from AERONET almucantar retrievals (Holben et al., 1998; Dubovik and King, 2000) are used. The SSA uncertainty is expected to be ±0.03 for AOD > 0.4 (Dubovik et al., 2000; 2002). Ideally, only quality assured Level 2.0 data should be used for all stations. However, the AOD threshold of 0.4 used in Level 2.0 product eliminates the bulk of the data and those left only represent the tail of data distribution, therefore, except for 10 stations that have sufficient amount of Level 2.0 data, Level 1.5 cloud screened data (Smirnov et al., 2000) is used for most stations. In section 4 I will discuss the rationale for making this data selection in detail. The expected accuracy of AERONET AOD is ±0.01.

The stations are selected purely based on the availability of an extensive data record. Specifically, the all-point measurements are first averaged into seasonal means. For each seasonal mean value to be reliable, I define the minimum number of observations required as 120 for Level 1.5 data and 60 for Level 2.0 data. For the data record at a station to be qualified for trend analysis, at least six data equivalent years is required. A total of 70 stations are selected, among which 10 are Level 2.0 stations and the rest are Level 1.5 stations. In addition, to better examine aerosol compositional information, the selected stations are categorized into five representative aerosols types, including biomass burning, dust, urban, rural and oceanic. The classification is based on previous studies (e.g., Kinne et al., 2003; Kahn et al., 2010; Garcia et al., 2012) and the description of the stations on the AERONET website (http://aeronet.gsfc.nasa.gov). Table 1 lists the information of the stations, and their geographic locations and aerosol types are indicated by Figure 1. The study focuses on the ten year period from 2003 to 2012, during which the measurements are most abundant and continuous and thus the sampling effect is minimal.

Table 1. Location, aerosol type, number of seasons used in the analysis (N) and 675 nm AOD, SSA and ABS trends at the 70 selected stations. Bold values indicate trends at 90% significance level

Station Name	Longitude	Latitude	Type	N	AOD Trend	SSA Trend	ABS Trend
East US							
Billerica	-71.269	42.528	Rural	37	0.0008	**0.068**	**-0.0049**
BONDVILLE	-88.372	40.053	Rural	36	0.0008	-0.015	0.0029
CARTEL	-71.931	45.379	Urban	41	-0.0023	**-0.15**	**0.014**
CCNY	-73.949	40.821	Urban	35	0.0054	**0.14**	-0.0020
Egbert	-79.75	44.226	Rural	41	**-0.013**	-0.11	**-0.013**
FORTH_CRETE	25.282	35.333	N/A[+]	34	**-0.020**	0.020	-0.0033
GSFC[*]	-76.84	38.992	Urban	24	-0.0045	**0.019**	**-0.0016**
Halifax	-63.594	44.638	Rural	41	-0.0011	0.021	-0.0008
MD_Science_Center	-76.617	39.283	Urban	40	**-0.016**	0.030	-0.0030
SERC	-76.500	38.883	Rural	35	-0.0011	**0.13**	**-0.011**
Wallops	-75.475	37.942	Rural	39	0.0018	-0.022	-0.0012
West US							
Bratts_Lake	-104.7	50.28	Rural	39	-0.0021	**0.087**	**-0.0033**
BSRN_BAO_Boulder	-105.006	40.045	Rural	41	**-0.0069**	0.041	**-0.0035**
Fresno	-119.773	36.782	Urban	34	**-0.022**	**0.067**	**-0.0094**
KONZA_EDC	-96.61	39.102	Rural	37	-0.0023	**-0.18**	**0.0098**
La_Jolla	-117.25	32.87	N/A	30	**-0.011**	0.014	**-0.0011**
Railroad_Valley	-115.962	38.504	Dust	42	**-0.0065**	**0.20**	**-0.0097**
Rimrock	-116.992	46.487	Rural	39	-0.0045	**0.087**	**-0.0058**

Table 1. (Continued)

Station Name	Longitude	Latitude	Type	N	AOD Trend	SSA Trend	ABS Trend
East US							
Saturn_Island	-123.133	48.783	Rural	29	0.0082	-0.033	**0.0047**
Sevilleta	-106.885	34.355	Rural	32	**0.0065**	-0.028	0.0050
Sioux_Falls	-96.626	43.736	Rural	32	**-0.017**	**0.090**	**-0.0075**
South America							
Alta_Floresta	-56.104	-9.871	Biomass	37	**-0.075**	**0.065**	**-0.015**
Arica	-70.313	-18.47226	Urban	26	**-0.043**	**-0.046**	**0.0048**
CEILAP-BA	-58.5	-34.567	Urban	40	0.0052	**0.092**	-0.0033
CUIABA-MIRANDA	-56.021	-15.729	Biomass	36	**-0.034**	**0.071**	-0.0051
IER_Cinzana[*]	-5.934	13.278	Biomass	29	0.014	-0.0041	0.0067
Rio_Branco	-67.869	-9.957	Biomass	31	**-0.043**	-0.039	0.0046
Sao_Paulo	-46.735	-23.561	Biomass	29	-0.017	**0.012**	**-0.022**
Europe							
Avignon	4.878	43.933	Urban	39	-0.0022	**0.101**	**-0.012**
Belsk	20.792	51.837	Rural	33	-0.0051	-0.014	0.0036
Carpentras	5.058	44.083	Urban	42	**-0.017**	**0.033**	**-0.0066**
Ispra	8.627	45.803	Urban	40	**-0.036**	**0.042**	**-0.0078**
IFT-Leipzig	12.435	51.352	Urban	32	-0.0017	-0.027	0.0034
Lecce_University	18.111	40.335	Urban	41	**-0.019**	**0.037**	**-0.0055**
Lille	3.142	50.612	Urban	42	**-0.019**	**0.063**	**-0.0089**
Minsk	27.601	53.92	Urban	40	-0.012	**0.13**	**-0.019**

Station Name	Longitude	Latitude	Type	N	AOD Trend	SSA Trend	ABS Trend
			Europe				
Missoula	-114.083	46.917	Rural	38	0.0022	**-0.056**	0.0023
Modena	10.945	44.632	Urban	31	**-0.044**	**0.071**	**-0.013**
Moldova	28.816	47.000	Urban	38	-0.0049	0.022	**-0.0040**
Moscow_MSU_MO	37.510	55.700	Urban	39	-0.017	**0.068**	**-0.011**
Oostende	2.925	51.225	Urban	28	**-0.036**	**0.045**	**-0.0086**
Palaiseau	2.208	48.700	Rural	41	-0.0065	**0.052**	**-0.0078**
Rome_Tor_Vergata	12.647	41.84	Urban	37	-0.0088	0.0042	-0.0032
Toravere	26.46	58.255	Rural	35	**-0.013**	**0.076**	**-0.0059**
Venise	12.508	45.314	Urban	42	**-0.023**	-0.016	-0.0002
			Africa				
Baniozoumbou[*]	2.665	13.541	Dust	34	-0.041	0.0088	0.00001
Capo_Verde	-22.935	16.733	Dust	42	**-0.21**	**0.042**	**-0.011**
Dakar[*]	-16.969	14.394	Dust	38	-0.021	**0.013**	**-0.0067**
Evora	-7.912	38.568	Rural	39	0.0070	-0.0094	0.0020
Ilorin[*]	4.34	8.32	Dust	20	-0.15	-0.0076	-0.0046
IMS-METU-ERDEMLI	34.255	36.565	Rural	38	-0.0024	0.012	-0.0019
Izana	-16.499	28.309	Rural	37	**-0.0070**	**0.023**	**-0.0013**
Nes_Ziona	34.789	31.922	N/A	33	-0.0002	0.016	-0.0008
SEDE_BOKER	34.782	30.855	Dust	40	0.013	-0.025	0.0039

Table 1. (Continued)

Station Name	Longitude	Latitude	Type	N	AOD Trend	SSA Trend	ABS Trend
Africa							
Solar_Village[*]	46.397	24.907	Dust	42	**0.080**	**0.0059**	0.0056
India							
Kanpur[*]	80.232	26.513	Dust	30	0.014	**0.020**	-0.0066
Northeast Asia							
Beijing[*]	116.381	39.997	Urban	38	**-0.10**	**0.029**	**-0.030**
Shirahama[*]	135.357	33.693	Rural	26	-0.035	**0.041**	**-0.019**
XiangHe[*]	116.962	39.754	Rural	32	-0.0654	-0.0070	0.0013
Australia							
Canberra	149.111	-35.271	Urban	42	-0.0007	**-0.16**	**0.0063**
Jabiru	132.893	-12.661	Biomass	37	**0.0087**	0.058	-0.0059
Lake_Argyle	128.749	-16.108	Biomass	41	0.0078	**-0.058**	0.0039
Others							
Ascension_Island	-14.415	-7.976	Oceanic	31	-0.012	0.025	0.0023
Bonanza_Creek	-148.316	64.743	Biomass	27	0.0099	**0.12**	**-0.012**
Dalanzadgad	104.419	43.577	Dust	35	**0.017**	**-0.092**	0.030
La_Parguera	-67.045	17.97	Oceanic	38	0.0089	**0.048**	-0.0010
Mauna_Loa	-155.578	19.539	Oceanic	42	**0.0012**	-0.0085	**0.0003**
Mexico_City	-99.182	19.334	Urban	33	**-0.038**	0.021	**-0.014**
Nauru	166.916	-0.521	Oceanic	41	**0.025**	**-0.11**	**0.0063**

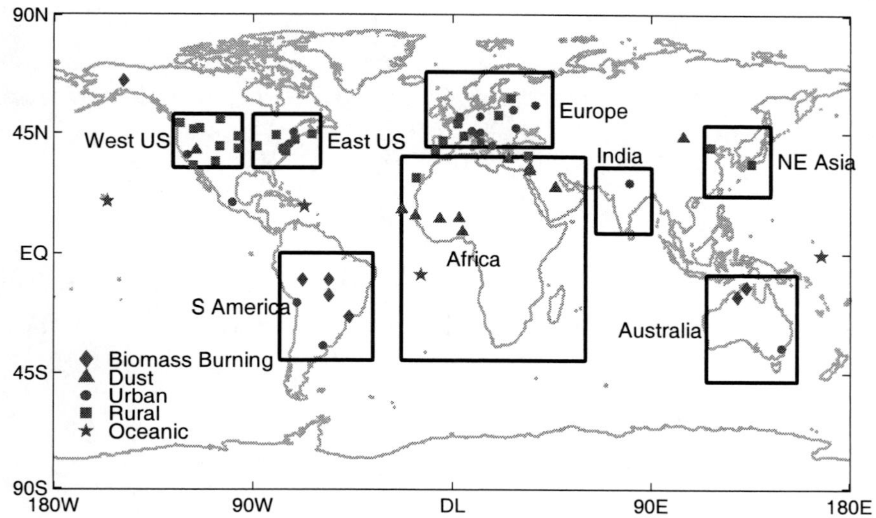

Figure 1. Location and aerosol types of the selected stations. The black boxes show the spatial definitions of the regions used in regional analysis.

Moreover, to minimize the influence of cases with high data variability or bias from sampling, the weighted linear regression by Yoon et al. (2012) is adopted here. This technique effectively reduces the influence of outliers on the estimated trend. Mathematical description of the method is given below.

The unweighted ordinary least square regression can be expressed as:

$$y_t = \beta_0 + \beta_1 x_t + \varepsilon_t$$

where y_t is the seasonal mean data at season x_t. The linear trend, β_1, and the constant β_0, are solved by minimizing the least square error:

$$S = \sum_{t=1}^{T} (y_t - \beta_0 - \beta_1 x_t)^2$$

It can be shown that the best linear unbiased estimator (BLUE) for $\beta = (\beta_0, \beta_1)$ is

$$\hat{\beta} = (X^T X)^{-1} X^T Y$$

where $X = (x_1, \cdots, x_T)$ and $Y = (y_1, \cdots, y_T)$.

The number of data samples for each season and outliers in the data both introduce uncertainties in the trend estimation. Therefore, the weighting factor is determined as

$$w_t = n_t / \sigma_t$$

where n_t is the number of observations for season x_t and σ_t is the standard deviation. The weighted linear regression problem then becomes minimizing

$$WS = \sum_{t=1}^{T} w_t (y_t - \beta_0 - \beta_1 x_t)^2$$

And the BLUE estimator becomes

$$\hat{\beta} = (X^T WX)^{-1}(X^T WY)$$

where $W = (w_1, \cdots, w_T)$.

The weighted regression is performed on the de-seasonalized time series constructed by removing the multi-year averaged seasonal cycle, in order to eliminate any impact from seasonal variability. Student's t test is used to test the significance level of the linear trend.

3. RESULTS

Linear trends are first estimated at each individual station for SSA, ABS and AOD. The maps showing the magnitude, as well as the significance levels of the trends of the three parameters at 440 nm, are presented in Figure 2. The numerical values of the trends are listed in Table 1. From the first panel of Figure 2, the majority of the global stations exhibit positive SSA trends, especially over North America, South America, and Europe, where trends are found to be as large as 0.1/decade. Out of the 70 stations, 33 have significant increasing SSA trends while only 8 have significant decreasing trends, at or above 90% significance level. Correspondingly, global ABS is dominated by a decreasing trend, with similar distribution as the SSA trend. In contrast, the trends in AOD are in general less evident or insignificant. Weak negative AOD trends are found over most regions except the Arabian Peninsular, where a slightly increasing trend is observed. The largest decrease in AOD is found over West Africa.

As mentioned above, it is observed that all three panels in Figure 2 display spatially coherent patterns. For example, North America, South America, Europe and East Asia all exhibit near uniform decreases in ABS, increases in SSA and slight decreases in the AOD. This provides the basis for the further examination of regionally averaged trends. The stations are grouped into eight regions, and their spatial definitions are presented in Figure 1. Due to limited knowledge of the spatial representativeness the stations, both global average (including all stations) and regional averages are calculated by assigning equal weight to each station. Figure 3 and Figure 4 show the global and regional mean, de-seasonalized SSA, ABS and AOD time series and trends, respectively. Only statistically significant trends are shown in Figure 3 and Figure 4. Consistent with Figure 2, global mean SSA exhibits a positive trend at ~0.03/decade, ABS has a decreasing trend at ~ -0.002/decade, while no significant trend is found in global mean AOD. Regionally, the largest positive SSA trend is found over South America, followed by West US, East US, Europe and Northeast Asia. These regions also have largest decreasing trends in ABS. While negative AOD trends are only found for East US, South America and Europe. Note that the effect of weighted regression is also well demonstrated by Figure 3 and Figure 4. For example, some large peaks, such as the 2004 one in the SSA time series for East US, does not heavily impact the linear trend due to its larger variance and thus smaller weight. Figure 5 summarizes the global and regional trends and their 90% significance levels. The first panel of Figure 5 clearly shows that all regions display positive SSA trends. Except for Africa and Australia, all trends are statistically significant. Similarly, a statistically significant negative ABS trend is found for all regions excluding Africa, India and Australia. The largest SSA and ABS trends are found over South America, followed by Europe, West US and NE Asia. With respect to AOD, except for the decreases over South America and Europe, no significant trend shows up for the other regions. According to the SSA, ABS and AOD trends shown by Figure 2 to Figure 5, it is highly plausible that decreases in aerosol absorption, rather than increases in aerosol scattering, is responsible for the observed changes in aerosol optical properties.

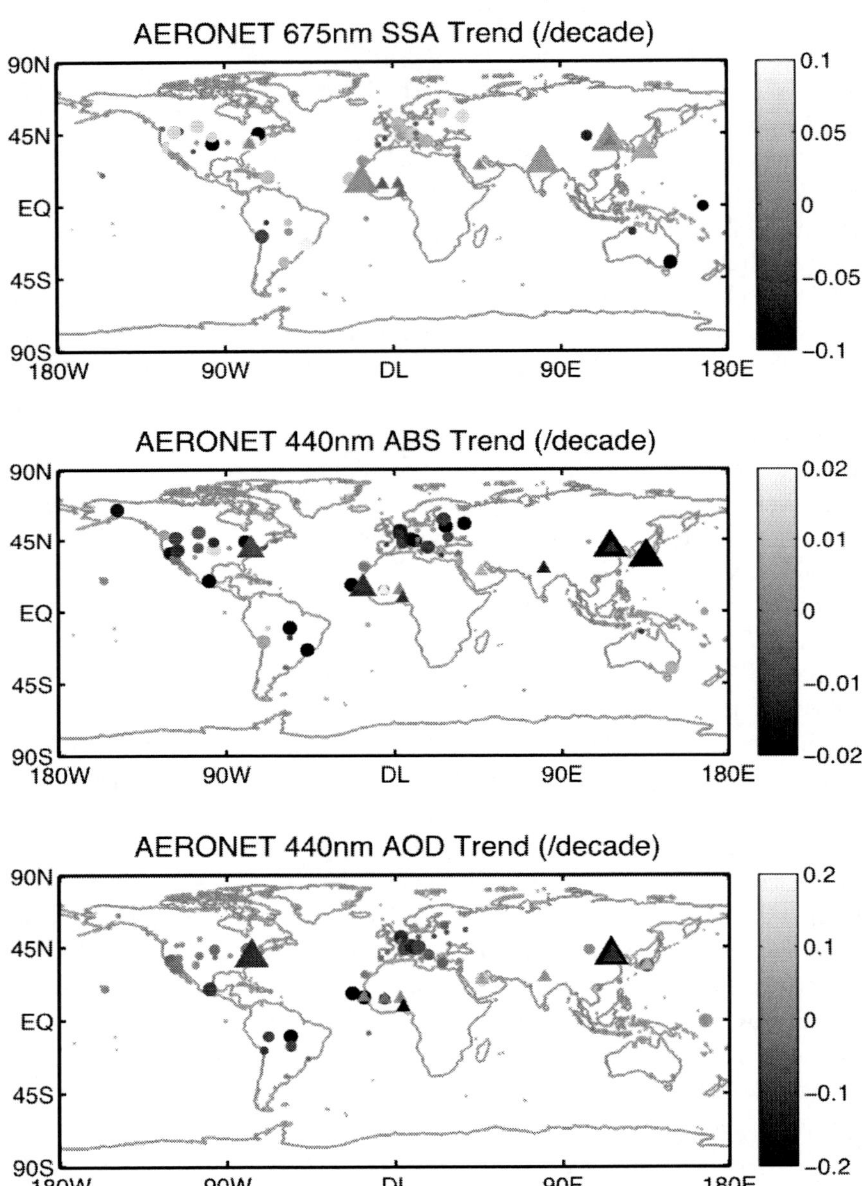

Figure 2. Magnitude and significance level of 440 nm SSA, ABS and AOD trends at the 70 selected stations. Triangles mean that Level 2.0 data are used for those stations. The size of the symbols denotes the significance level, ranging from: Large, 99%; Medium, 95%; Small, 90%; Tiny, <90%.

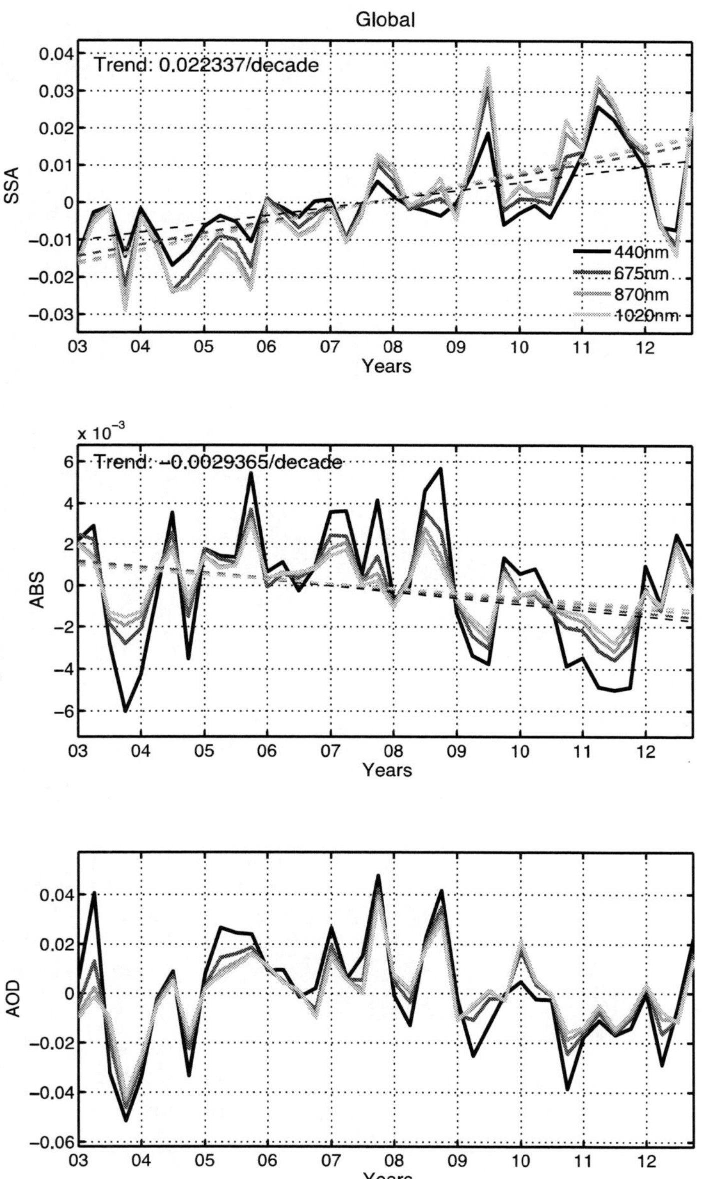

Figure 3. Globally averaged de-seasonalized SSA (upper), ABS (middle) and AOD (lower) time series and trends. AOD does not have significant trends so no trend line is shown.

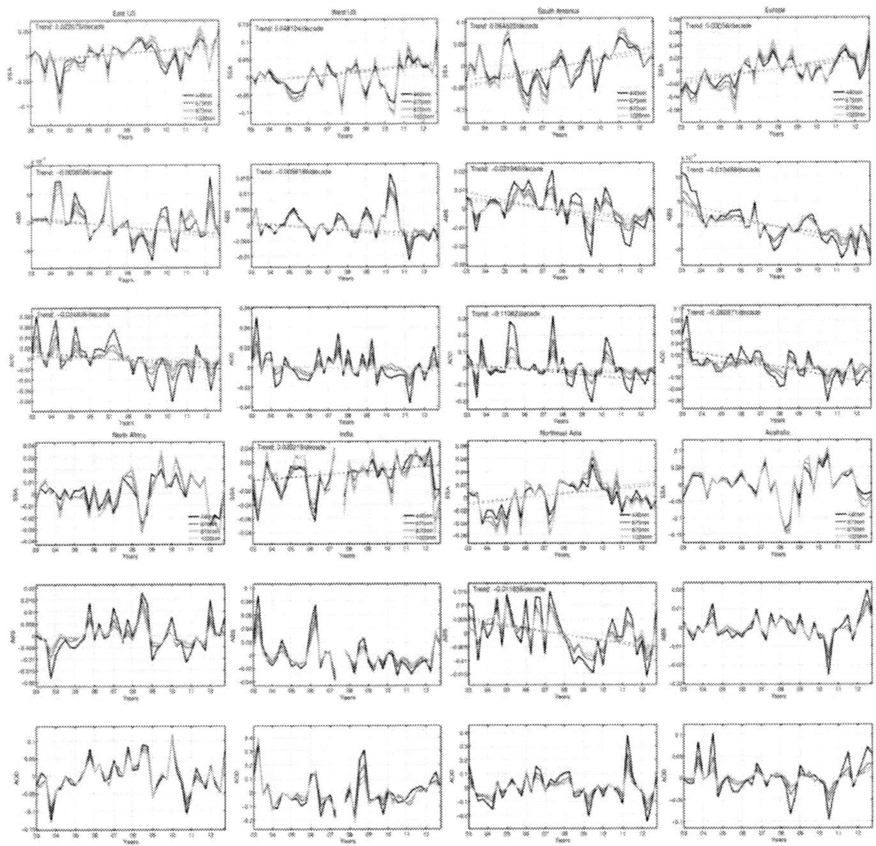

Figure 4. Time series and trends for SSA, ABS and AOD averaged over the 8 regions defined in Figure 1. Only statistically significant trends are shown.

It is also necessary to examine the trends averaged from stations representing different aerosol types, so that aerosol compositional changes can be better inferred. The time series and trends for the five representative aerosol types are shown in Figure 6 and Figure 7. It is found that except for dust sites, biomass burning, urban, rural and oceanic sites all exhibit significant positive trends in SSA and negative trends in ABS. As black carbon should be the principal absorber for these four aerosol types, this result is a further indication of black carbon decrease.

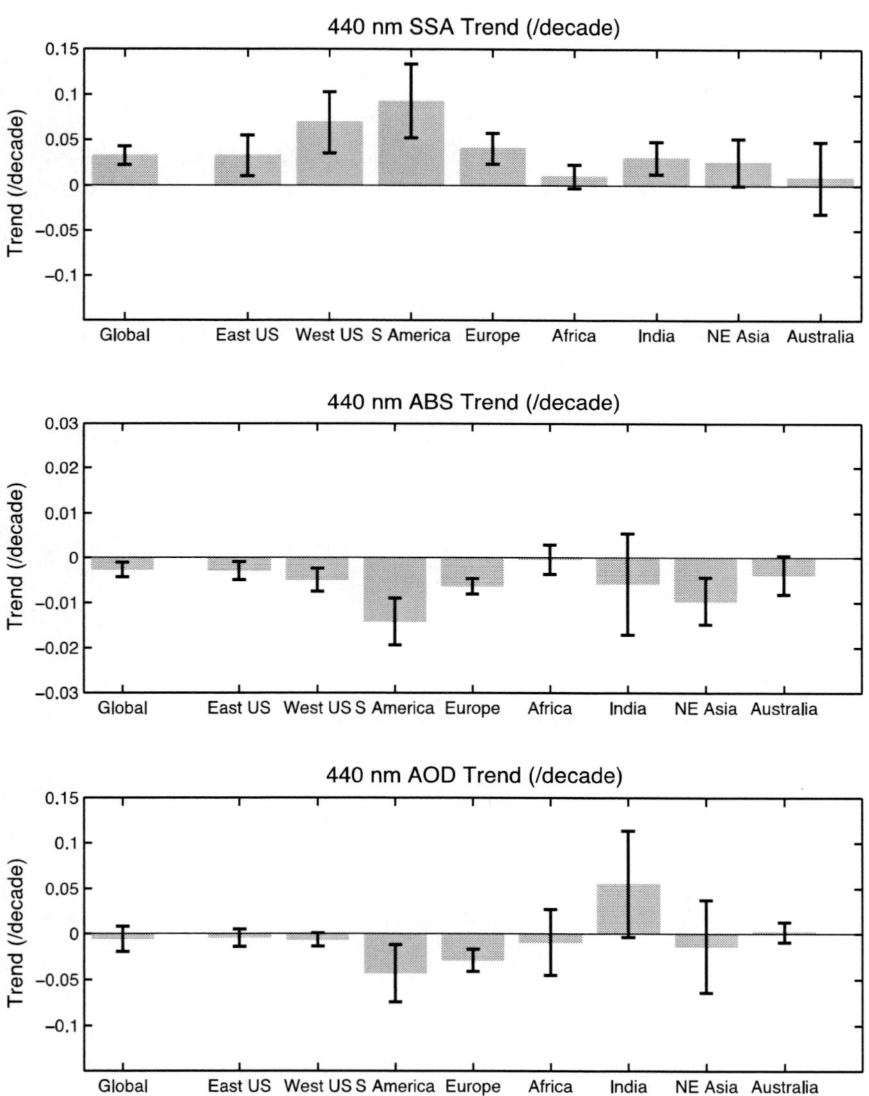

Figure 5. Global and regional 440 nm SSA, ABS and AOD trends. The error bars indicate the upper and lower limits of the 90% confidence interval of the estimated trends.

                                    Jing Li

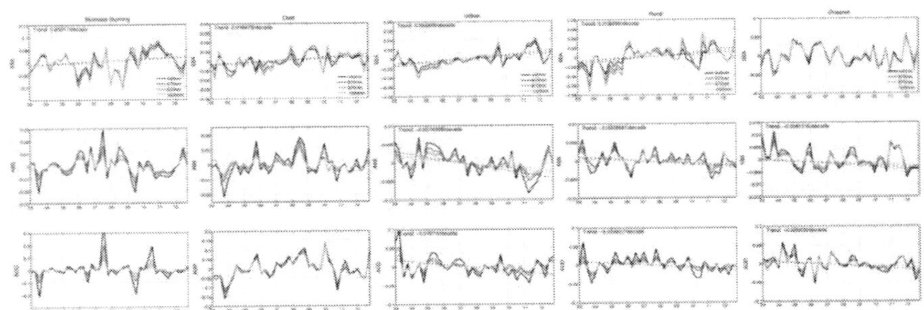

Figure 6. Averaged SSA, ABS and AOD time series and trends for the five representative aerosol types. Only statistically significant trends are shown.

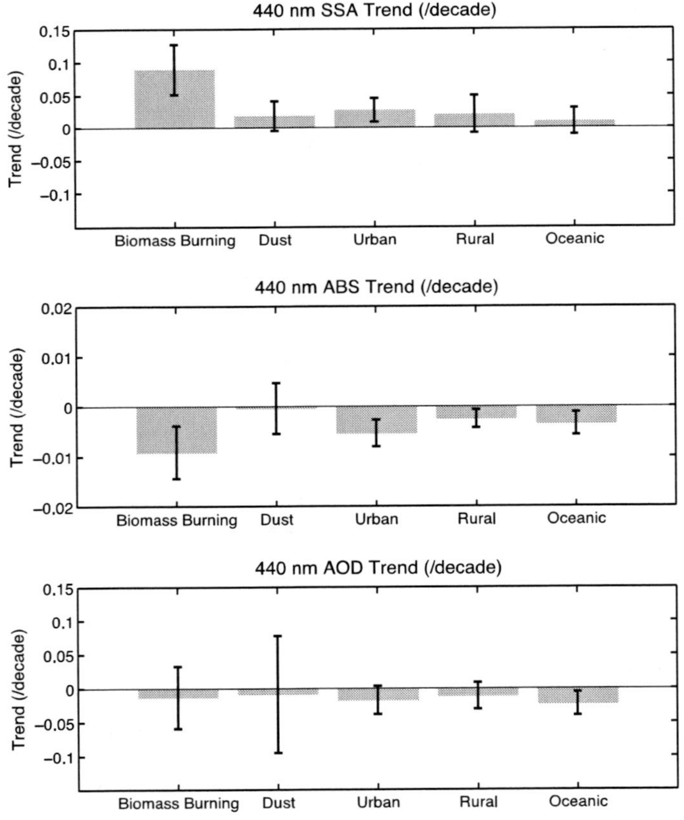

Figure 7. Trends at 440 nm for the five aerosol types. The error bars indicate the upper and lower limits of the 90% confidence interval of the estimated trends.

Indeed, the spectral dependence of SSA has different shapes for aerosol masses dominated by black carbon absorption and dust absorption. This feature is clearly demonstrated by Figure 8, which shows averaged SSA spectral dependence for the different regions (Figure 8a) and different aerosol types (Figure 8b), respectively. From Figure 8b, we can see that dust aerosols have a different spectral shape from the other types, with SSA increase with wavelength. This is due to stronger absorption by mineral dust at UV and near-UV wavelengths than at longer wavelengths, which results in a lower scattering/extinction ratio and thus lower SSA value at short wavelengths. In contrast, the absorption spectral for black carbon is relatively flat, and since the extinction optical depth generally decreases with wavelength, the fraction of absorption in total extinction becomes larger as wavelengths increases. As a result, SSA decreases with wavelength. Figure 8a also indicates that the spectral shape of SSA for dust-influenced regions, including North Africa and India, and NE Asia (mostly during the spring) are different from the other regions, where the absorbing species is primarily black carbon. The different spectral behaviors of dust and black carbon aerosols can be quantified as the Absorption Angstrom Exponent (AAE), with is defined as the spectral dependence of aerosol absorption

$$AAE = \frac{\log(\tau_{abs\_w1} / \tau_{abs\_w2})}{\log(w_1 / w_2)}$$

where τ_{abs_w1} and τ_{abs_w2} are absorption optical depth at the two wavelengths w_1 and w_2, respectively. Here the two wavelengths are chosen as 440 nm and 870 nm.

The AAE parameter is closely related to aerosol composition, and can be used to characterize different aerosol mixtures (Bergstrom et al., 2007; Russell et al., 2010). The theoretical value for the AAE is 1 for pure black carbon, and increases with mixture of organic carbon and dust (Russell et al., 2010). However, the AAE parameter alone usually does not provide clear separations of different aerosol types or mixtures. Giles et al., (2012) added another piece of information, the Angstrom Exponent (AE) parameter which is a measure of averaged aerosol size, and performed cluster analysis of AAE and AE. Their results showed that these two parameters combined could effectively constrain aerosol type, as dust aerosols generally have much larger sizes. In particular, the signature of black carbon and dust dominated aerosol masses are characterized as two distinct AE-AAE clusters. In this study, as we are mostly interested in the changes of aerosol property, I take advantage of Giles et al. (2012) results and perform an EOF analysis on the AE-AAE cluster to examine the temporal evolution of aerosol composition. In this part of the analysis, only Level 2.0 data is used for all 70 stations. The reason is that the variability in the Level 1.5 data is much higher with occasional outliers. The wide spread of the data will seriously contaminate the signals, or the center of the AE-AAE cluster, and will prevent the EOF analysis to extract meaningful patterns. However, in the previous estimation of linear trends, the noise in Level 1.5 data has been taken into account by the weighted regression.

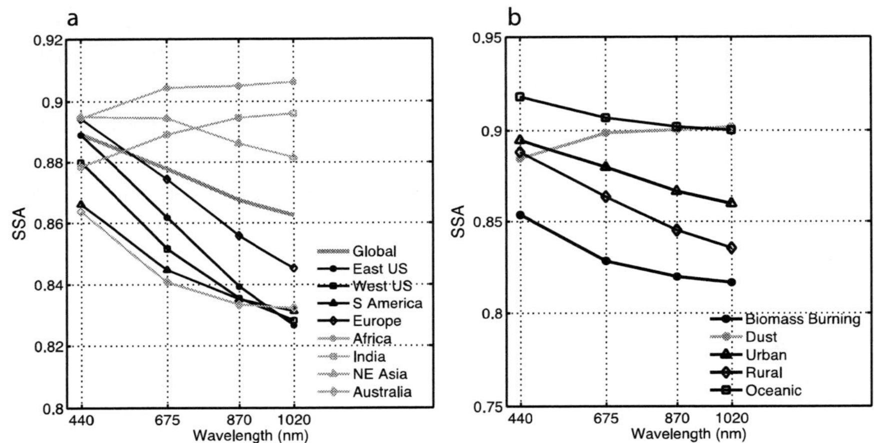

Figure 8. Averaged spectral dependence of SSA for the regions (a) and aerosol types (b). It is clearly seen that dust and dust dominant regions have different spectral shapes, with SSA increasing with wavelength.

The following two paragraphs describe the analysis procedure in detail. First, the AE-AAE clusters for each season are obtained by calculating the Joint Density Distribution (JFD) field of these two parameters at 50×50 resolution. The range considered for the AE and AAE are 0-2.5 and 0-3, respectively. This results in a data matrix that has dimension 50×50×40, where 40 denotes the number of seasons (4 seasons per year for 10 years). The mean JFD field averaged over the entire study period is shown in Figure 9. The three well-separated clusters correspond to dust, mixed and urban/industrial-biomass burning aerosols respectively according to Giles et al. (2012). Note that the urban/industrial and biomass burning aerosols are not fully separated, because black carbon is dominant absorbing species for both types, which results in the similarity of their clusters.

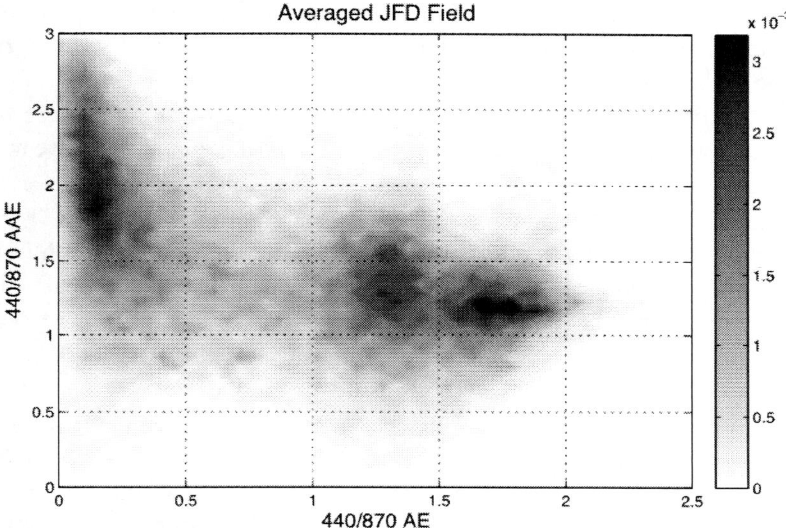

Figure 9. Averaged Joint Frequency Distribution (JFD) field of AE and AAE. The left cluster should correspond to dust aerosol, the right one corresponds to urban and biomass burning, while the middle one indicates mixed aerosol types, according to Giles et al. (2012).

Next, the data matrix is decomposed by EOF analysis to extract spatial and temporal variability. It is found that the spatial pattern of the second EOF mode (left panel of Figure 9) exhibits an AE-AAE distribution that highly resembles the urban-industrial and biomass burning aerosol clusters shown in Figure 7c & d of Giles et al. (2012). The PC time series of this mode displays a statistically significant decreasing trend (right panel of Figure 10). Since the absorbing species associated with this cluster is primarily black carbon (AAE ~= 1 and AE > 1.5), the spatial pattern and time series combined suggest a decreasing trend in global black carbon fraction during the past ten years.

In a final an attempt to eliminate the effect of dust, the AE and AAE values in the EOF Mode 2 cluster is used to sort the data for the two dust-dominate regions, North Africa and India, in order to further extract the trend for black carbon. Specifically, only ABS, SSA and AOD data at conditions satisfying $1 \leq AAE \leq 1.5$ and $1.5 \leq AE \leq 2$ are selected to construct the new time series. The newly estimated ABS, SSA and AOD trends using sorted data for

North Africa are now more consistent with the other regions (Figure 11), with statistically significant negative trends in ABS and AOD and positive trends in SSA. This result is a further support of a worldwide decrease in aerosol absorption due to black carbon. For India, due to the limited amount of data left after the sorting, the estimation of the trend is not possible so the figure is not shown here.

To summarize, a global reduction in aerosol absorption optical depth, associated with an increase in aerosol single scattering albedo but a weak trend in the extinction optical depth are found by analyzing AERONET Level 1.5 and Level 2.0 inversion products. The EOF analysis of the AE-AAE JFD field further implies that this trend is likely the consequence of a reduction in black carbon over worldwide locations. This result has significant impact on our estimation of aerosol climate effects, as increases in SSA implies more cooling effect from aerosols to counteract the warming effect of greenhouse gases. Moreover, reduction in black carbon results in the indirect effect of aerosols, i.e., the interaction of aerosols with the hydrological cycle. This may further shift aerosol forcing to the cooling end, although the indirect effect is much less well understood. Another impact of trends in aerosol absorption is on satellite retrievals of aerosol properties. As mentioned earlier, satellite retrieval algorithms assume constant aerosol absorption or SSA values. However, an increase of SSA will lead to an underestimation of their assume SSA value and possibly a high bias in the retrieve aerosol trends.

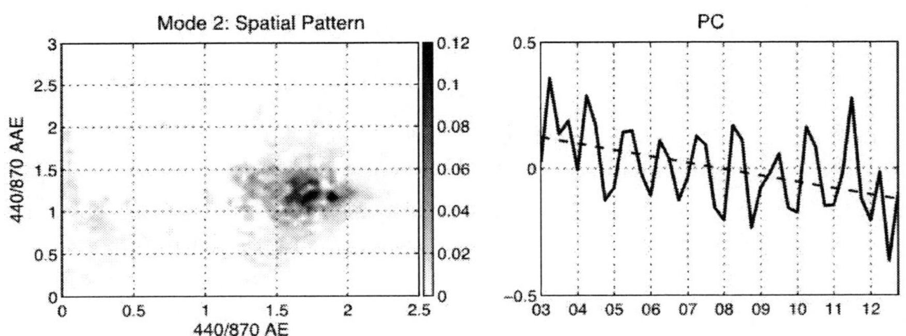

Figure 10. The second Mode of EOF analysis of the JFD field. The analysis is performed using Level 2.0 data only for all 70 stations. The spatial pattern of should correspond to black carbon, with AE > 1.5 and AAE close to 1. This mode is associated with a decreasing time series, which suggests global black carbon fraction may have decreased during the past 10 years.

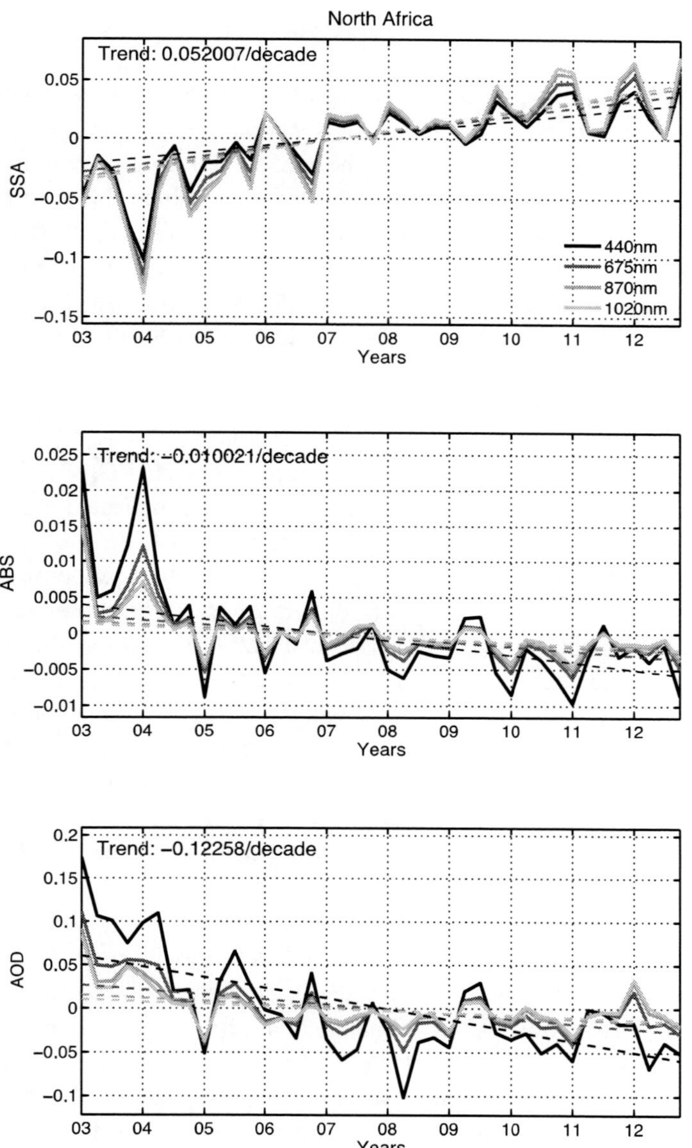

Figure 11. Trends of North Africa stations using data sorted by the AE-AAE cluster shown in Figure 8. Now the trends are consistent with most of the other stations, showing positive SSA and negative ABS trends. Since this cluster mainly captures black carbon absorption, this result is a further support for black carbon reduction.

4. DISCUSSIONS

In the last section, I present an increasing SSA trend, associated with a decreasing ABS trend over worldwide locations, which can be attributed to black carbon reduction. Admittedly, the AERONET data used in this study is not perfect. The Level 1.5 data that are used in weighted linear regression for most stations have even higher uncertainties. However, an important reason for using Level 1.5 data is that the bulk of the data will be excluded by the quality assurance criteria defined by the AERONET team which requires AOD at 440 nm is greater than 0.4 (Holben et al., 2006). More importantly, the re-sampling induced by quality control will produce a serious bias towards dust and will miss the information of the other aerosol types. Next, I will discuss this in detail.

Figure 12 shows the distribution (histogram) of AERONET AOD at 440 nm for the five aerosol types. As previous studies indicated that the AOD distribution is most appropriately modeled as lognormal (Ignatov et al., 2000), a fitted lognormal probability density function (pdf) is also superimpose on top of the histograms. The small panel on the upper right corner of each plot shows the distribution of AOD between 0 and 0.5. Clearly, the bulk of the AODs are concentrated at the small end for all aerosol types. As a result, we can infer that the minimum threshold value of 0.4 only captures the tail of the AOD distribution. In fact, according to the fitted lognormal distribution, the probability that AOD > 0.4 are 13.47%, 31.89%, 18.28%, 10.02%, and 2.71%, for biomass burning, dust, urban, rural and oceanic sites, respectively. Restricting to only high AOD cases will inevitably miss the bulk of the information.

Most importantly, the information on aerosol composition will also change if selecting data at different AOD thresholds. In order to demonstrate this effect, data are selected from the 1st to the 4th quantile of the fitted lognormal distribution of AOD and their SSA spectral dependence is compared. The results are shown in Figure 13. It can be clearly seen that the SSA spectral shapes in the highest quantile (gray triangle curves) are biased towards dust, with a much flatter or even reversed spectral slope of SSA compared to the lower quantiles. Although biomass burning, urban and oceanic stations are not dominated by dust, they all more or less have dust intrusions. And for the dust stations, some located at the Sahel region also have seasonal influences of biomass burning aerosols. Figure 13 indicates that by selecting only the high AOD cases, our aerosol sample will be biased towards dust.

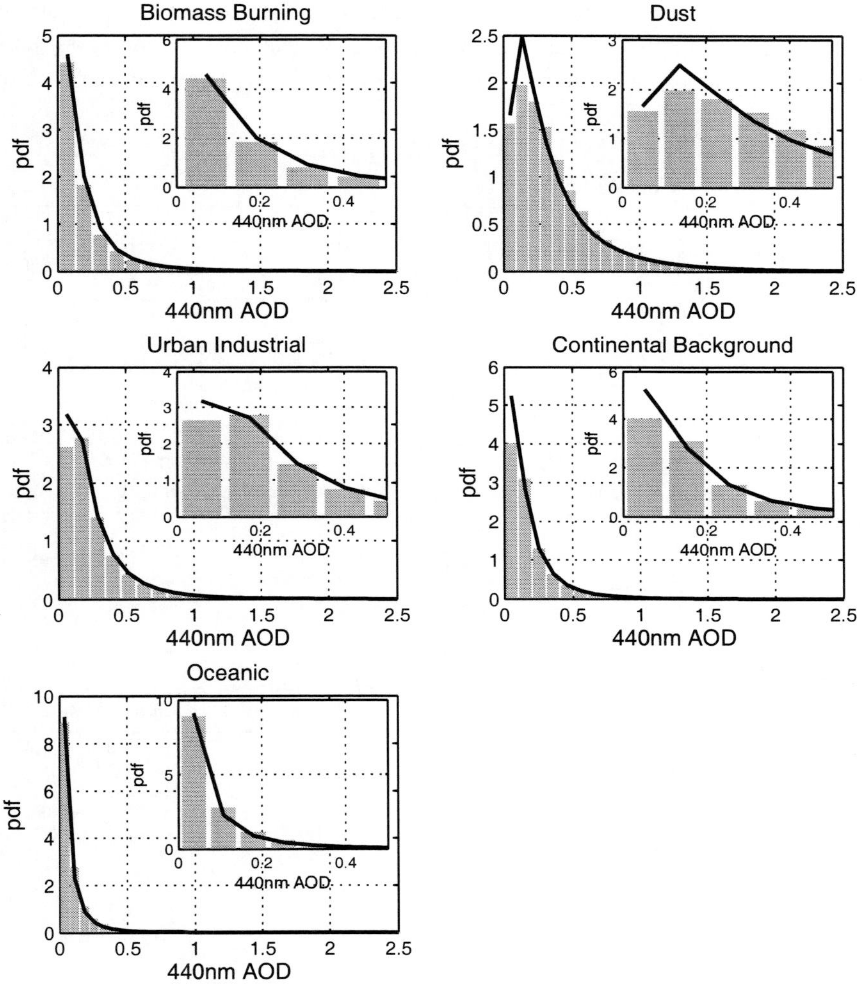

Figure 12. The histogram and fitted lognormal (black curves) distribution of AOD at 440 nm for the five aerosol types. The small panel on the upper right shows the distribution at small AOD values, from 0 to 0.5. It is seen that AOD > 0.4 only represents the tail of the distribution.

The distribution of AOD and the change of SSA spectral shape with it suggest that the small end of AOD is not negligible for complete knowledge of all aerosol types, despite the higher uncertainty of the data. On one hand, the weighted least square regression applied here mitigates the noise issue and greatly reduces the potential bias of the trends by large errors and outliers. On the other hand, trends reported here are consistent with previous model studies for Europe and US (Novakov et al., 2003; Streets and Aunan, 2005) and independence measurements for Japan (Kudo et al., 2010), which increase our confidence in the Level 1.5 data.

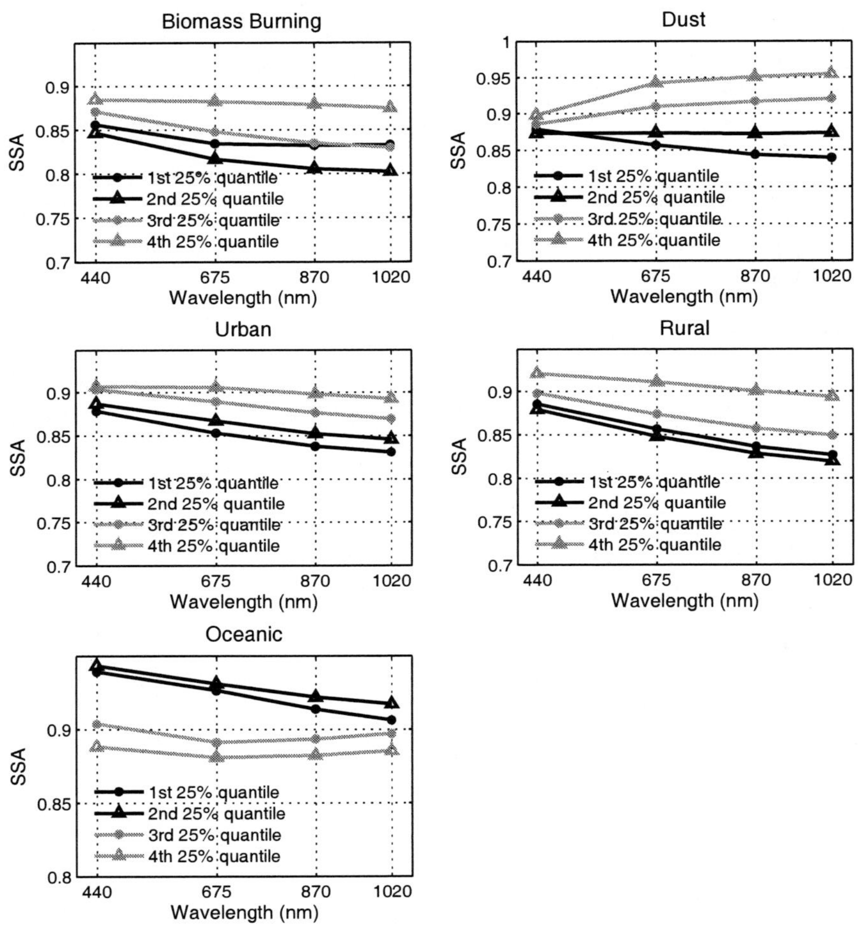

Figure 13. Spectral dependence of SSA selected by the 1st to 4th quantile of the AOD distribution. Except for rural sites, the spectral SSA at the last quantile (probability > 75%) for biomass burning, dust, urban and oceanic sites all show the dominance of dust. This is an indication that by selecting only large AOD cases, the aerosol sample will be biased towards dust.

CONCLUSION

In this study, a near globally uniform increase in aerosol SSA and decrease in ABS over the past ten years is reported, as revealed by AERONET measurements. Weak trends in the AOD suggests the increase of SSA is directly associated with the decrease in ABS. By analyzing the temporal evolution of the AE-AAE clusters using the EOF technique, it is found that trends can be attributed to the decrease in the fraction of black carbon aerosols.

The reduction in black carbon and the associated ABS and SSA trends over Europe and US are consistent with previous modeling studies (Novakov et al., 2003; Streets and Aunan, 2005). The trend in Japan is corroborated by independent measurements by Kudo et al. (2010). The upward SSA trend over these regions could also play an important role in observed global "brightening" (e.g., Norris and Wild, 2007), which has been previously attributed to AOD reduction. The trend over China and India seems controversial to black carbon emission trend estimates for these two countries (e.g., Streets et al., 2006; Ohara et al., 2007; Lu et al., 2011). However, qualified stations are limited over these two countries and the data quality may not be as good, e.g., the two China stations show opposite trends and the ABS trend for India are not significant. Additional data or longer records are likely required to confirm the AERONET trends for China and India.

The results presented here imply that the warming due to black carbon may have weakened and the net aerosols radiative forcing may have become more negative. This could be a potential factor in the "global warming pause" since 1998 (e.g., Kaufmann et al., 2011; Guemas et al., 2013). Another important consequence of trends in SSA is the impact on satellite retrieved trends of aerosol properties. As most satellite retrieval algorithm assume constant SSA, systematic changes in SSA over time may produce spurious tendencies in the retrieved AOD (Mishchenko et al., 2007). Lyapustin et al. (2012) indicated that the temporal change in the bias between MODIS and AERONET AOD over Beijing can be explained by the increase of SSA. Mishchenko et al. (2012) showed that in AVHRR AOD retrieval, decreasing SSA by 0.07 from the start to the end of the data record could eliminate the downward trend in the data.

Nonetheless, there are still limitations in the AERONET data. Most of the stations are established in recent years, making their records too short for trend analysis. Different temporal samplings for different stations also bring uncertainties to the regional trends. It also lacks the ability to directly retrieve aerosol composition. However, AERONET is currently the most extensive surface network that provides ABS and SSA measurements. AERONET data have been widely used to validate satellite retrievals and the information has been used as inputs to satellite retrieval algorithms. As trends in SSA is an important issue in both climate change and satellite retrieval, this study calls for the expansion of the network both spatially and temporally, as well as the development of more advanced sensors to monitor detailed aerosol properties.

ACKNOWLEDGEMENTS

The author thanks the AERONET team for providing the data set used in this study.

REFERENCES

Bergstrom, R. W., Pilewskie P., Russell P. B., Redemann J., Bond T. C., Quinn P. K., and Sierau B. Spectral absorption properties of atmospheric aerosols, *Atmos. Chem. Phys.* 2007, 7, 5937–5943.

Dubovik, O. and King M. D. A flexible inversion algorithm for retrieval of aerosol optical properties from Sun and sky radiance measurements, *J. Geophys. Res.* 2000, 105, 20 673-20 696.

Dubovik, O., Smirnov A., Holben B. N., King M. D., Kaufman Y. J., Eck T. F., and Slutsker I. Accuracy assessments of aerosol optical properties retrieved from Aerosol Robotic Network (AERONET) Sun and sky radiance measurements, *J. Geophys. Res.* 2000, 105, 9791–9806.

Dubovik, O., Holben B. N., Eck T. F., Smirnov A., Kaufman Y. J., King M. D., Tanré D., and Slutsker I. Variability of Absorption and Optical Properties of Key Aerosol Types Observed in Worldwide Locations. *J. Atmos. Sci.* 2002, 59, 590–608.

García, O. E., Díaz J. P., Expósito F. J., Díaz A. M., Dubovik O., Derimian Y., and Roger J. C. Shortwave radiative forcing and efficiency of key aerosol types using AERONET data. *Atmos. Chem. Phys.* 2012, 12, 5129-5145.

Giles, D. M., Holben B. N., Eck T. F., Sinyuk A., Smirnov A., Slutsker I., Dickerson R. R., Thompson A. M., and Schafer J. S. An analysis of AERONET aerosol absorption properties and classifications representative of aerosol source regions, *J. Geophys. Res.* 2012, 117, D17203, doi:10.1029/2012JD018127.

Guemas V., Doblas-Reyes F. J., Andreu-Burillo I., and Asif M. Retrospective prediction of the global warming slowdown in the past decade. *Nature Climate Change* 2013, doi:10.1038/nclimate1863

Hansen, J., Sato M., and Ruedy R. Radiative forcing and climate response. J. Geophys. Res. 1997,102, 6831-6864.

Holben, B.N., Eck T. F., Slutsker I., Tanré D., Buis J. P., Setzer A., Vermote E., Reagan J. A., Kaufman Y. J., Nakajima T., Lavenu F., Jankowiak I. and Smirnov A. AERONET-A federated instrument network and data archive for aerosol characterization, *Remote Sens. Environ.* 1998, 66, 1-16.

Holben B. N., Eck T. F., Slutsker I., Smirnov A., Sinyuk A. et al. AERONET's Version 2.0 quality assurance criteria", *Proc. SPIE* 6408, Remote Sensing of the Atmosphere and Clouds, 64080Q (November 28, 2006); doi:10.1117/12.706524.

Ignatov, A., Holben B. N., and Eck T. F. The lognormal distribution as a reference for reporting aerosol optical depth statistics; Empirical tests using multi-year, multi-site AERONET Sunphotometer data. *Geophys. Res. Lett.* 2000, 27, 3333-3336.

Kahn, R. A., Gaitley B. J., Garay M. J., Diner D. J., Eck T. F., Smirnov A., and Holben B. N. Multiangle Imaging SpectroRadiometer global aerosol product assessment by comparison with the Aerosol Robotic Network. *J. Geophys. Res.* 2010, *115*(D23).

Kaufmann, R. K., Kauppib H., Mann M. L. and Stock J. H. Reconciling anthropogenic climate change with observed temperature 1998–2008. *Proc. Natl Acad. Sci. USA* 2011, 108, 790–793.

Kinne, S., Lohmann U., Feichter J., Schulz M., Timmreck C., Ghan S., and Kaufman Y. J. Monthly averages of aerosol properties: A global comparison among models, satellite data, and AERONET ground data. *J. Geophys. Res.* 2003, *108*(D20), 4634.

Kudo, R., Uchiyama A., Yamazaki A., Sakami T., and Kobayashi E. From solar radiation measurements to optical properties: 1998–2008 trends in Japan, *Geophys. Res. Lett.* 2010, 37, L04805, doi:10.1029/2009GL041794.

Lyapustin, A., et al. Reduction of aerosol absorption in Beijing since 2007 from MODIS and AERONET, *Geophys. Res. Lett.* 2011, 38, L10803, doi:10.1029/2011GL047306.

Mishchenko, M. I., et al. Accurate monitoring of terrestrial aerosols and total solar irradiance: Introducing the Glory mission. *Bull. Amer. Meteor. Soc.* 2007, 88, 677–691.

Mishchenko, M.I., Liu L., Geogdzhayev I. V., Li J., Carlson B. E., Lacis A. A., Cairns B., and Travis L. D. Aerosol retrievals from channel-1 and -2 AVHRR radiances: Long-term trends updated and revisited., J. Quant. Spectrosc. Radiat. Transfer 2012, 113, 1974-1980, doi:10.1016/ j.jqsrt.2012.05.006.

Norris, J.R. and Wild M. Trends in aerosol radiative effects over Europe inferred from observed cloud cover, solar "dimming", and solar "brightening", *J. Geophys. Res.* 2007, 112, D08214.

Novakov, T., Ramanathan V., Hansen J. E., Kirchstetter T. W., Sato M., Sinton J. E., and Sathaye J. A. Large historical changes of fossil-fuel black carbon aerosols, *Geophys. Res. Lett.* 2003, 30(6), 1324, doi:10.1029/2002GL016345.

Ohara, T., Akimoto H., Kurokawa J., Horii H., Yamaji K., Yan X., and Hayasaka T. An Asian emission inventory of anthropogenic emission sources for the period 1980-2020, *Atmos. Chem. Phys.* 2007, 7, 4419-4444.

Ramanathan, V., Crutzen P. J., Kiehl T. J. and Rosenfeld D. Aerosols, Climate, and the Hydrological Cycle, *Science* 2001, 294, 2119-2124.

Smirnov, A., Holben B. N., Eck T. F., Dubovik O., and Slutsker I. Cloud-screening and quality control algorithms for the AERONET database. *Remote Sensing of Environment* 2000, 73(3), 337-349.

Streets, D. G., and Aunan K. The importance of China's household sector for black carbon emissions, *Geophys. Res. Lett.* 2005, 32, L12708, doi:10.1029/2005GL022960.

Yoon, J., von Hoyningen-Huene W., Kokhanovsky A. A., Vountas M., and Burrows, J. P. Trend analysis of aerosol optical thickness and Ångström exponent derived from the global AERONET spectral observations, *Atmos. Meas. Tech.* 2012, 5, 1271-1299, doi:10.5194/amt-5-1271-2012.

In: Aerosols
Editor: Adam Yanick Pearson

ISBN: 978-1-63117-512-1
© 2014 Nova Science Publishers, Inc.

Chapter 2

LATITUDINAL DISTRIBUTION PATTERN OF BERYLLIUM-7 IN NEAR-SURFACE ATMOSPHERIC AEROSOLS IN THE EAST ASIAN MONSOON CLIMATE SYSTEM, CHINA AND IMPLICATIONS ON AIR POLLUTANTS

Yong-Liang Yang[*], Xiao-Hua Zhu, Nan Gai,
Jing Pan and Ke-Yan Tan
National Research Center of Geoanalysis, Beijing, China

ABSTRACT

The activities of the cosmogenic nuclide beryllium-7 (^{7}Be) in near-surface atmospheric aerosols simultaneously collected at five cities in the East Asian monsoon region during four seasons were measured using a high-volume air sampler and a high-resolution gamma-ray spectrometer. The latitudinal distribution of the annual average ^{7}Be concentrations in aerosols follows a normal distribution pattern, with the maximum (8.31±0.49 mBq m^{-3}) occurring at ~ 40°N, which was significantly higher than values reported for other cities in the East Asian monsoon region and in the world during the same period. The simultaneous

[*] Corresponding Author address: National Research Center of Geoanalysis, 26 Baiwanzhuang Avenue, Xicheng District, Beijing 100037, China, Corresponding author's email: ylyang2003@hotmail.com.

observation of ^{7}Be at different latitude sampling sites in the East Asian monsoon region also displayed approximately a normal distribution pattern, but with the maximum at 30°N in spring and autumn and at 40°N in summer and winter. At a particular time, the latitudinal distribution pattern can instantaneously move southward or northward depending on season. Higher mean ^{7}Be concentrations in spring and autumn could be attributed to the transition of atmospheric circulation patterns. On the other hand, near-surface atmospheric ^{7}Be concentrations were lower in winter and summer with northward shift of the peak position. Atmospheric circulation may also play an important role in the atmospheric long-range transport (LRT) of persistent organic pollutants (POPs), but the extent to which it affects the latitudinal distributions of POPs in atmosphere and surface soils remains uncertain. Using ^{7}Be as a reference, simultaneous observation of ^{7}Be and typical POP compounds at different latitude in the East Asian monsoon climate system (EAMCS) showed similar latitudinal distribution patterns, suggesting that atmospheric circulation can affect the latitudinal distributions of POPs in the atmosphere.

Keywords: Atmospheric aerosol, beryllium-7, latitudinal distribution, East Asian monsoon, China

INTRODUCTION

Due to the irregularity of emission of air pollutants varying both in time and space and randomness in meteorological changes, the geographic distribution of air pollutants in local areas on a short-term time scale often shows a random feature. Thus we need a reference or tracer to elucidate the effect of climate on the pollutants' geographic distribution in atmosphere and soils in a particular region. The tracer should have no sources from land and its geographic distribution in troposphere is mainly governed by atmospheric circulation.

Beryllium-7 (^{7}Be) is a natural radionuclide originated by interaction of cosmic rays with nitrogen and oxygen in the stratosphere and upper troposphere. The seasonal and longitudinal changes in ^{7}Be atmospheric production rate can be neglected (Lal et al., 1958). It changes with altitude and latitude but in a relatively stable way. When ^{7}Be is formed, it quickly binds with atmospheric aerosols. In the top upper troposphere, its production rate is two-fold higher than that in near-surface (Yoshimori, 2005). The half-life of ^{7}Be is 53.3 d and the residence time in the troposphere is about 35 d

(Bleichrodt, 1978; Dutkiewicz and Husain, 1985). ^{7}Be can be transported from stratosphere, which serves as the source for ^{7}Be production and accumulation, into the troposphere via stratosphere-to-troposphere exchange (STE) or through the Brewer-Dobson circulation on a global scale (Feely et al., 1989; Holton et al., 1995). The average concentration of ^{7}Be in atmospheric aerosols observed at stations with altitude >700 m are reported as 3.54 mBq m^{-3}, while the average concentration at low altitude was 1.84 mBq m^{-3} with the global mean concentration of ^{7}Be in atmospheric aerosols as 2.45 mBq m^{-3} (Kulan et al., 2006). It was found that ^{7}Be concentrations followed a pattern similar to that of the fallout fission products (Todorovic et al., 2010) which may enter the stratosphere during nuclear weapons tests or accidents of nuclear power plants.

Due to the influence of geomagnetism and atmospheric circulations, the distribution of ^{7}Be in atmosphere relies on latitude. It is well known that annual mean ^{7}Be concentrations in near-surface aerosols are related to both of the latitude and altitude of the sampling site (Bourcier et al., 2011). Kulan et al. (2006) measured ^{7}Be at 61 locations, mostly in Europe, ranging from Greenland to the South Pole and observed that the annual average ^{7}Be concentrations follows a normal distribution pattern, with the maximum (5.5 mBq m^{-3}) occurring at ~40°N. As the atmospheric circulation pattern in Europe is different from the East Asian monsoon system, therefore it is necessary to investigate the latitudinal distribution of atmospheric ^{7}Be in the East Asian monsoon zone.

Beryllium-7 can be an ideal tracer for atmospheric circulations and aerosols on a short-term time scale. Beryllium-7 has no source on the Earth surface and its concentrations in the atmosphere are not affected by human activities. It can be transported by atmospheric circulations and deposited through dry and wet precipitation to the ground. Once deposited, it will not return to the atmosphere. Several studies on ^{7}Be concentrations in atmospheric aerosols have been reported. Lee et al. (2004) reported airborne ^{7}Be concentrations in Hong Kong using back-trajectories. A few studies on ^{7}Be in the air of China were mainly focused on the western part of China, such as Qinghai-Tibetan Plateau and Mt. Qomolangma (Mt. Everest) region (Lin et al., 2003; Zheng et al., 2005; Wan et al., 2006) which are in the domain of South Asian monsoon and in high altitude regions. Most of reports on ^{7}Be in the atmospheric environment in the East Asian monsoon region were focused on scavenging processes such as wet and dry precipitation of ^{7}Be (Momoshima et al., 2006; Hasegawa et al., 2007; Akata et al., 2008; Shi et al., 2011).

With advantage of short half-life and easy determination, ^{7}Be has been widely used as a tracer in atmosphere science. Most studies on application of ^{7}Be as a tracer in atmosphere were restricted in aerosol itself, such as estimation of the mean residence time of aerosols (Koch et al., 1996; Ayub et al., 2009; Ali et al., 2010). A few researches were involved in relations with ozone and some air pollutants in the atmosphere such as nitrate, sulfate, and mercury (Igarashi et al., 1998; Lamborg et al., 2000; Garimella et al., 2003). There is little report available on simultaneous study on ^{7}Be in the atmospheric aerosols at different latitude and simultaneous study on ^{7}Be and air pollutants in the atmospheric aerosols at different latitude.

The long-range transport process of air pollutants including nitrogen oxides (NOx), sulfur dioxide (SO$_2$) (Elliot et al., 1997; Van Aardenne et al., 1999), heavy metals, mercury, and persistent organic pollutants (POPs) has been a crucial issue for environmental concerns in countries downwind of the significant emissions. Persistent organic pollutants (POPs) are a group of toxic compounds characterized with lipophilic and semi-volatile properties. They are chemically stable, persistent in environment and can be bio-magnified via food chain, with chronic toxicity, mutation and carcinogenicity (Hansen, 2000). POPs become the global environmental issue. The United Nation Environment Protection (UNEP) has established the global treaty known as the Stockholm Convention on elimination of production of POPs and controlling unintentionally produced POPs (UNEP, 2001), among which organochlorine pesticides (OCPs), polychlorinated biphenyls (PCBs), and dioxins (PCDD/Fs) are mostly concerned. These compounds can undergo long-range transport (LRT) in form of vapor or adsorbed onto atmospheric aerosols to remote clean areas via atmospheric circulation at regional or even global scales (Wania and Mackay, 1993) and may deposit onto the surface media through dry and wet precipitation or air-soil exchange, posing potential threat to human health and ecosystem in remote areas. The more volatile components of the mixtures, such as the lower chlorinated PCBs, hop poleward more efficiently than the higher chlorinated ones, leading to a compositional shift to more volatile constituents with increasing northern latitude (Ockenden et al., 1998; 2003).

The concepts of "cold condensation" and global fractionation" have been formulated in interpretation of the mechanisms of semi-volatile organic pollutants migrating to polar areas through evaporation and atmospheric transport (Wania and Mackay, 1996; Ockenden et al., 1998, 2003; Gouin et al., 2004). However, only a small fraction of the inventory of POPs has reached the far north in the North Hemisphere. Soils have had a key role in 'protecting' the Arctic, by retarding re-cycling and retaining POPs, particularly in the major C stores of temperate forests, grasslands and peat bogs (Ockenden et al.,

2003). Three factors are likely to have been influential in causing this observation, i.e., (i) (soil organic matter), (ii) climatic factor; (iii) emissions (Meijer et al., 2003). Here we proposed a mechanism for supplementary interpretation of latitude distribution pattern of POPs in East Asian monsoon climate system (EAMCS) using the cosmogenic nuclide ^{7}Be as a reference.

Atmospheric circulation can plays an important role in the atmospheric long-range transport of POPs, but the degree to which it affects the latitudinal distributions of POPs in atmosphere and soils is unclear. According to atmospheric circulation theory (Aguado and Burt, 2010), air near the equatorial regions receives more heat than in other regions, leading to air up-movement and move towards higher latitude. Due to the geostrophic force, westward wind velocity can be built up. The higher latitude the air mass moves towards, the greater the westward velocity and smaller the polar-ward velocity can be observed. When the air jet in the Hadley Cell from equatorial upwelling air meets the Ferrel Cell at the mid-latitude subtropical region, a horizontal convergence can be formed and sinking of air occurs (Fig. 1). Although the Hadley Cell and the Ferrel Cell both exist in different seasons, the two cells in North Hemisphere both move northward in summer or southward in winter in accordance with the solar radiation strength in different seasons (Zeng et al., 2011). In a particular period, the position of the sinking of air shows a south-north movement, and changes in their strength, depending on the movement of the air convergence.

The eastern China is located within the largest monsoon domain (the East Asian monsoon climate system, EAMCS) in the world, characterized by dominant warm and wet southeasterly in summer, and dry, cold northwesterly in winter covering North and Northeast China, Korean Peninsular and Japan (Aguado and Burt, 2010) (Fig. 2). Variations in meteorological fields associated with the monsoon can influence transportation, deposition, and chemical reactions of aerosols over the eastern China (Zhang et al., 2010). Besides, the West Pacific Subtropical High moves westwards in summer, strongly affect the meteorological system in East Asian continent, and moved back to the Western Pacific in winter season. The air mass from Siberian region during the formation of the Siberian anticyclone can intensify the stratosphere-to-troposphere exchange (STE)(Zhang et al., 2010) and strongly influence the weather in the East Asian monsoon zone during autumn, winter and early spring. As the concentration levels of pollutants in the atmosphere are affected by monsoons, the East Asian monsoon zone is an ideal place for study of ^{7}Be as a tracer of atmospheric long-range transport of air pollutants.

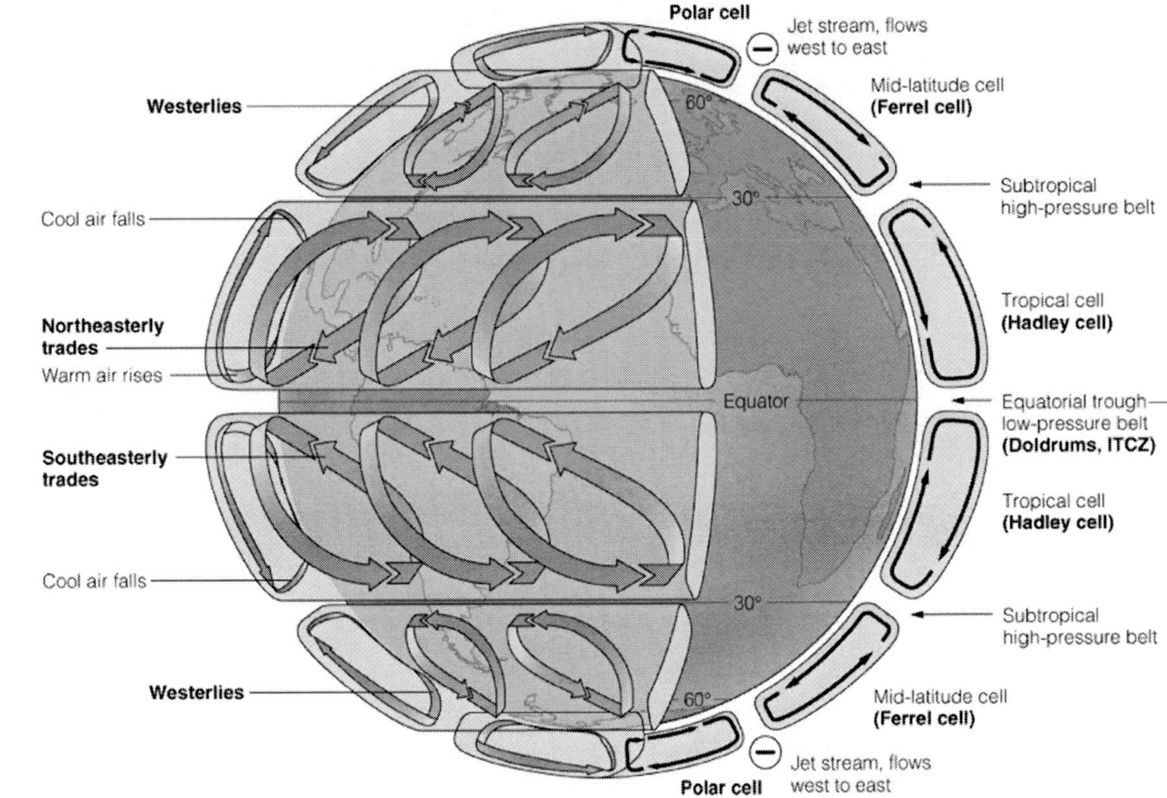

Figure 1. Schematic illustration of movement of air in atmospheric circulation cells (quoted from http://science.kennesaw.edu/~jdirnber/oceanography/LecuturesOceanogr/LecCurrents/LecCurrents.html).

In this chapter, the results of annual and short-term simultaneous measurement of ^{7}Be in near-surface atmospheric aerosols at different latitude cities in East Asian monsoon zone in China are discussed. Five cities were chosen as the investigation site in this study (Fig. 2). This study may provide base-line data for short-term behaviors of ^{7}Be in the atmosphere in aerosols in East Asian monsoon region during different seasons. The aim of this paper is to use the cosmogenic nuclide ^{7}Be as a reference to show that the atmospheric circulation in East Asian monsoon climate system (EAMCS) can link ^{7}Be in atmospheric aerosol at different latitude and affect the latitudinal distributions of POPs, and to further explore ^{7}Be as the geochemical tracer and provide a different perspective on the global fate of aerosols and air pollutants.

METHODS

Sample Collection

The sampling sites for measurement of ^{7}Be in aerosols were described in Table 1 and shown in Fig. 2. The sampling sites for consecutive measurement of ^{7}Be, OCPs, and PCBs in aerosols on weekly time scale were in Xicheng District of Beijing (39°54′ N, 116°24′ E), Laoshan District of Qingdao (35°36′ N, 119°30′ E; 75 m above sea level), and Tianhe District of Guangzhou (23°08.25′ N, 113°19′ E; about 25 m above sea level). Near-surface atmospheric aerosol samples were collected at the frequency of 3 d per week for one year.

Samples for simultaneous measurement of ^{7}Be, OCPs, and PCBs in near-surface atmospheric aerosols were collected at five cities in different seasons during 2012 and 2013. The sampling duration was one filter for each 24 h. Aerosol samples were collected by using a large volume TSP (total suspended particles) sampler (KC-1000 with a constant flow rate of 1.04 m^3 min^{-1}, Qingdao Laoshan Electronics Co, China) with a rectangular glass fiber filter film (GFF: 200 mm × 250 mm, pore size of 0.4 μm, 47 mm diameter, Waterman. USA). Prior to sampling, GFFs were baked at 450°C for 6 h to remove any organic contaminant. Air was collected at 10 m above the ground. The integrated sampling volume of each sample was calculated automatically by the sampler from the real flow rate and sampling time.

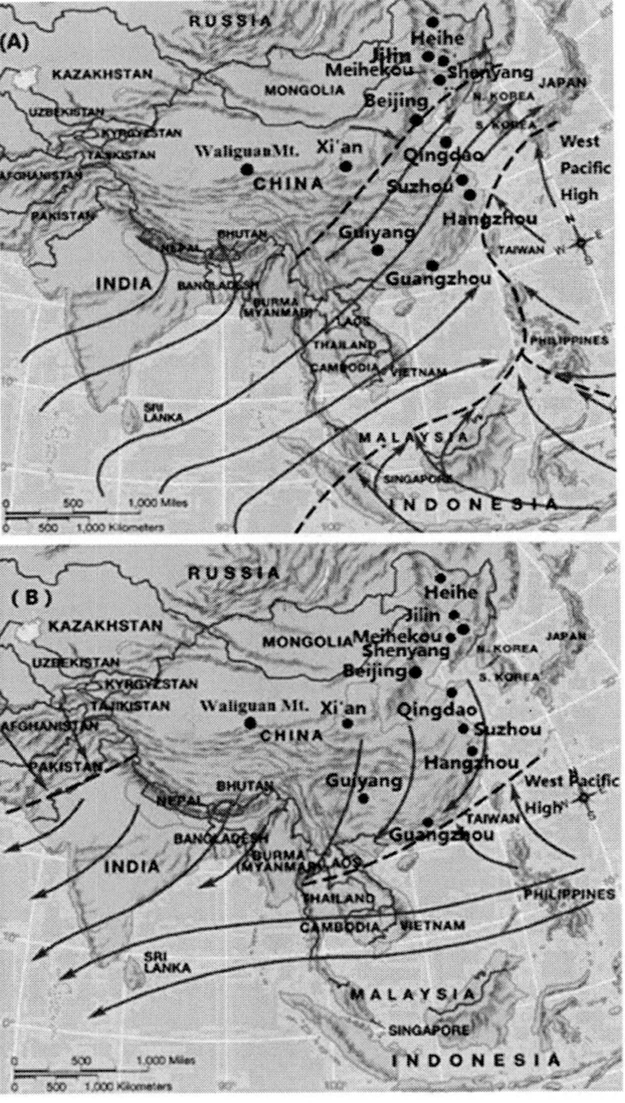

Figure 2. East Asian monsoon in (A) summer and (B) winter (modified from Aguado and Burt, 2010) and locations of the cities at different latitude in East Asian monsoon zone discussed in this chapter.

The samples for POP compounds in vapor form were collected by using XAD-2 resin (Supelpak^{TM}–2SV, 13673-U, SUPELCO, USA) based passive air samplers (PAS). Five sampling sites in Qingdao, Beijing, Shenyang,

Meihekou, and Heihe (Fig. 2) were installed with parallel samplers at each site. The averages of persistent organic pollutants in air were determined in October 2009-April 2010. In this study, the PAS sampler was based on and calibrated by the method of Toronto University (Wania et al., 2003). XAD-2 filled mesh cylinders were sealed in airtight containers with Teflon lids to avoid contamination during transit to the sites. Prior to deployment, XAD-2 resin was Soxhlet extracted using in turn methanol, acetonitrile, and dichloromethane (DCM). The XAD-2 resin (60 mL of wet XAD-2 in methanol) was transferred to a pre-cleaned stainless steel mesh cylinder and dried in a clean desiccator. Harvested XAD-2 cylinders were stored in sealed, solvent rinsed, glass jars stored at -20°C until extraction.

Table 1. Sampling Sites for Measurement of ^{7}Be in Near-surface Atmospheric Aerosols.

City	Location	Altitude (m)*	Longitude	Latitude
Heihe	Changfa Village	120	127.53 9°E	50.22 9°N
Beijing	Xicheng District	31.3	116°24' E	39°54' N
Qingdao	Laoshan District	75	119°30' E	35°36' N
Suzhou	Canglang District	5.0	120.58°E	31.30°N
Guangzhou	Tianhe District	25	113°19' E	23°08.25' N

*meters above sea level

Laboratory Procedures

(1) Beryllium-7

The glass–fiber filter sample was folded into a rectangle cube and then wrapped with aluminum foil with the bottom area of 64 mm × 52 mm. It was placed in a measuring vial, so that all filter papers had the same irradiation geometry to the germanium detector for ^{7}Be measurements. The ^{7}Be activity was measured by its 477.16 keV (the branching ratio: 10.5%) γ-ray using the high-resolution gamma-ray spectrometer measurement system (Canberra, US) with a BE5030 detector which has a γ-ray energy range of 3 keV ~ 3 MeV.

The measuring system is shielded with Canberra 747E lead chamber with the wall thickness of 10 cm. The detection efficiency was 50.93%. The spectrometer is operated by *Genie 2000* software and *DSA-1000* (Canberra, US) that allows automatic peak-search, calculation of peak area, reduction of background, self-absorption correction, decay correction, and interference correction. The counting time for ^{7}Be was > 43200 s and the counting error was <7 %. Laboratory Sourceless Calibration Software (LabSOCS, http://www.canbera.com/ products/839. asp) was used in establishment of the counting efficiency curve and treatment of ^{7}Be counting data obtained from gamma-ray spectrometry measurement for atmospheric aerosol samples, and calculation of the ^{7}Be activity (Bq) which was corrected automatically by the software of the gamma-ray spectrometer for the sampling time, sampling time length, the time interval between sampling and measurement, and measurement time interval to calculate the ^{7}Be specific activities (mBq m^{-3}) in the atmospheric aerosol sample. Decay correction was made to the date of sample collection.

(2) POPs

The extraction and cleanup procedure for OCPs and PCBs were carried out as follows. The GFF filter samples were added surrogate standards 2,4,5,6-tetrachloro-*m*- xylene (TMX) and PCB209, sohxleted with 150 mL DCM for 48 h. Prior to the extraction, 2 g copper chips were added in order to remove sulfur. The extract was rota-evaporated (≤35°C) to 5~10 mL, and added 10~15 mL hexane, rota-evaporated to 1 mL. This step was repeated three times and then the sample solution was transferred to a 5-mL cell bottle, and passed through a silica-alumina chromatograph column (7 mm i.d. made of glass), prepared by adding, from bottom to top, 10 g 3% activated silica, 10 g 3% de-activated alumina, and 1 g dehydrated sodium sulfate). The column was eluted with 35 mL Hexane/DCM (volume ratio = 1:1) solution. The effluent was rota-evaporated to 0.5 mL, then concentrated with gentle high purity nitrogen gas to 0.2 mL, and added 4 μL 5 μg mL^{-1} pentachlornitrobenzine (PCNB) as the internal standard, ready for GC analysis.

The analyses of OCPs and PCBs in all samples and field blanks were carried out by a gas chromatograph (HP6890) equipped with a ^{63}Ni electric capture detector (ECD), using a fused silica capillary column (HP-5MS, 30 m × 0.32 mm i.d.; with film thickness of 0.25 μm). The temperature settings for the injector and detector were 280°C and 320°C, respectively. The initial column temperature of 50°C was held for 2 min, then raised at a rate of 10°C min^{-1} to 180°C, held for 2 min, and raised to 220°C at 2°C min^{-1}, and to 290°C at 10°C min^{-1}, held for 15 min. Oxygen-free nitrogen (99.999% purity) was used as the carrier gas at a constant flow rate of 1.5 mL min^{-1}. The pre-column pressure was 103.45 kPa (15.0 psi). The sample (1.0 μL) was injected with splitless injection mode. The make-up gas was high pure nitrogen. GC peak identification of the sample was conducted by comparing the retention times of each analyte peak for the sample and the standard, and GC/MS (Finnigan Trace 2000) was utilized to identify unknown peaks.

Quality Control

The quality assurance and control of ^{7}Be data in our laboratory has been certified by taking part in the international interlab for radioactivity measurement of environmental samples organized by the International Atomic Energy Agency (IAEA) (Li et al., 2011).

Sampling blank, reagent blank, and procedure blank were tested during the POPs analysis, among which there were no target compounds in the reagent and procedure blanks. One Blank sample was carried out during each batch (ten samples) of experiment to ensure the cleanness of reagents and laboratory wares. Duplicate analyses were carried out for each batch of experiment to ensure the repeatability of the analytical results. The recoveries for OCPs were in a range of 61%-102%, and for PCB209 were 50%-105%. All the data were corrected for recovery. Through matrix spike experiment, the LOQ (limit of quantification) for OCPs was 0.04-0.1 pg m^{-3}, for PCBs was 0.36-0.71 pg m^{-3}.

RESULTS AND DISCUSSIONS

Annual Average Concentrations of ^{7}Be in EAMCS in China

The ^{7}Be concentrations measured on a weekly basis for one year in Beijing, Qingdao, and Guangzhou are shown in Fig. 3. The seasonal distribution patterns of ^{7}Be concentrations in Beijing, and Qingdao were in the sequence of autumn > spring > winter > summer, significantly different from that observed in the low latitude city Guangzhou which was in the sequence of spring > winter > summer > autumn. The annual average concentration of ^{7}Be in atmospheric aerosols in Beijing was 8.19±0.57 mBq m^{-3} (Tan et al., 2013). The annual average concentration of ^{7}Be in atmospheric aerosols at Qingdao showed 6.88±0.41 mBq m^{-3} (Yang et al., 2013). At the low latitude city Guangzhou, the annual average ^{7}Be was the lowest, only 2.59±0.31 mBq m^{-3}

(Pan et al., 2011). The concentrations of [7]Be in atmospheric aerosols in Beijing and Qingdao were significantly higher than the global average value of 2.45 mBq m^{-3}, and also higher than the global low altitude average value 1.84 mBq m^{-3} (Kulan et al., 2006).

The long-term distribution of the cosmogenic isotope [7]Be in surface air in Europe, extending from latitude 47°N to 68°N has been reported by Kulan et al. (2006), suggesting a decrease in average [7]Be activity with increasing latitude (Fig. 4). They have plotted [7]Be activity data obtained from stations worldwide, located at elevations $<$500 m in the northern hemisphere versus latitude and fitted their data with a Gaussian function as

$$F(x) = 2*10^3 + 3*10^3 \exp\{-0.5\,[(x-38.7)/8.4]^2\} \tag{1}$$

with a correlation coefficient R=0.99. As the most data summarized in their work are located north of 40°N (mainly in Sweden), most of the data points fall on the right side of the Gaussian curve. As the atmospheric circulation pattern in Europe is different from the East Asian monsoon system, therefore it is necessary to investigate the latitudinal distribution of atmospheric [7]Be in the East Asian monsoon zone.

The annual mean [7]Be concentrations measured in the authors' previous work (Pan et al., 2011; Tan et al., 2013; Yang et al., 2013) and data from literature (Wan et al., 2006; Jiang, 1994; Chang et al., 2008) are presented in Table 2 and their latitudinal distribution pattern are shown in Figure 3. The range of the annual mean concentrations was 2.57±0.31~8.39±0.49 mBq m^{-3}, with the maximum occurred at Beijing and the minimum at Guangzhou. Although some of these data were not measured in the same years, the fundamental feature of the curve will not be seriously changed due to the fact that the variations of [7]Be production rate in the atmosphere caused by solar activity variation are only within a range of ±20% (Kulan et al., 2006). It can be seen that the annual average near-surface atmospheric [7]Be activity between 20°N~50°N in the East Asian monsoon zone in China follows a Gaussian distribution-like pattern, with the highest value (8.31±0.57 mBq m^{-3}) occurring at the middle latitude station in Beijing.

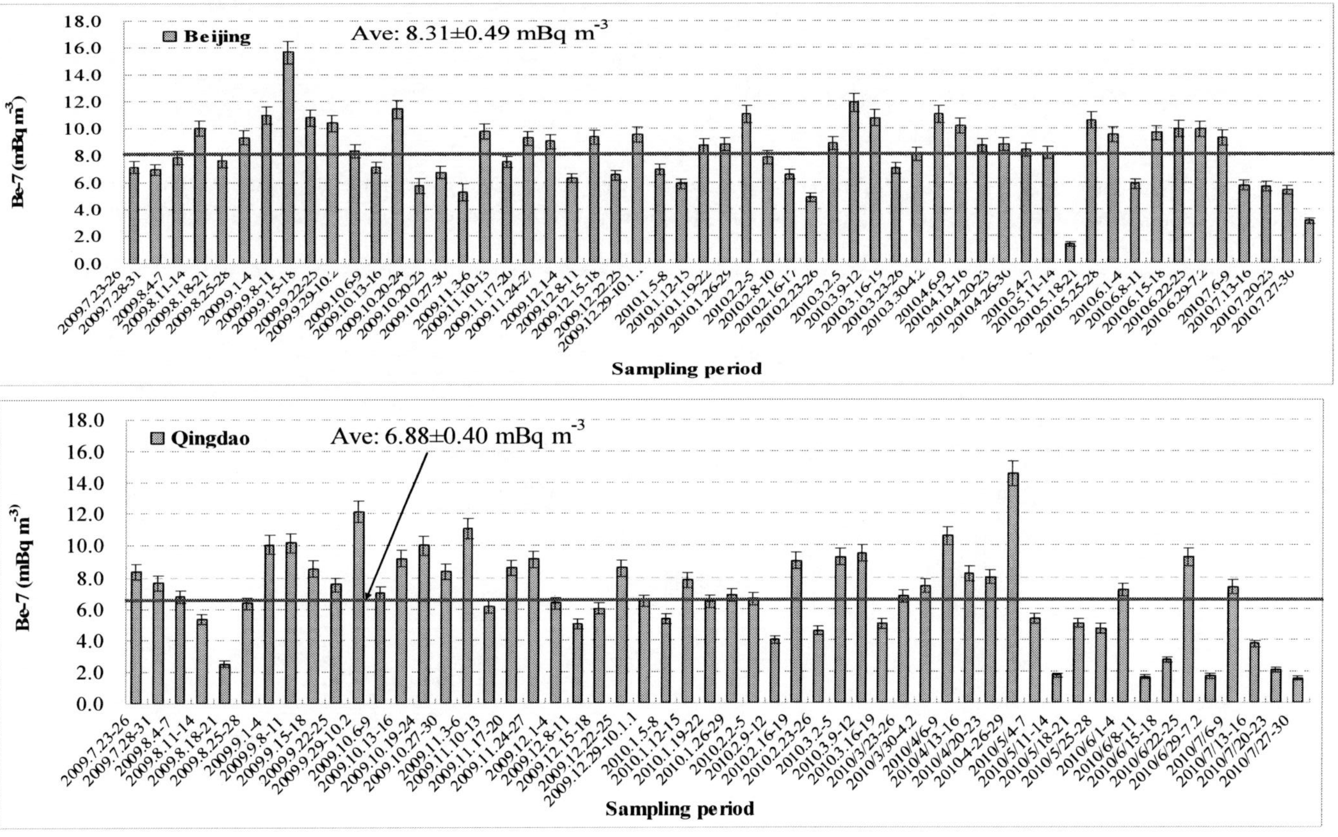

Figure 3. (Continued)

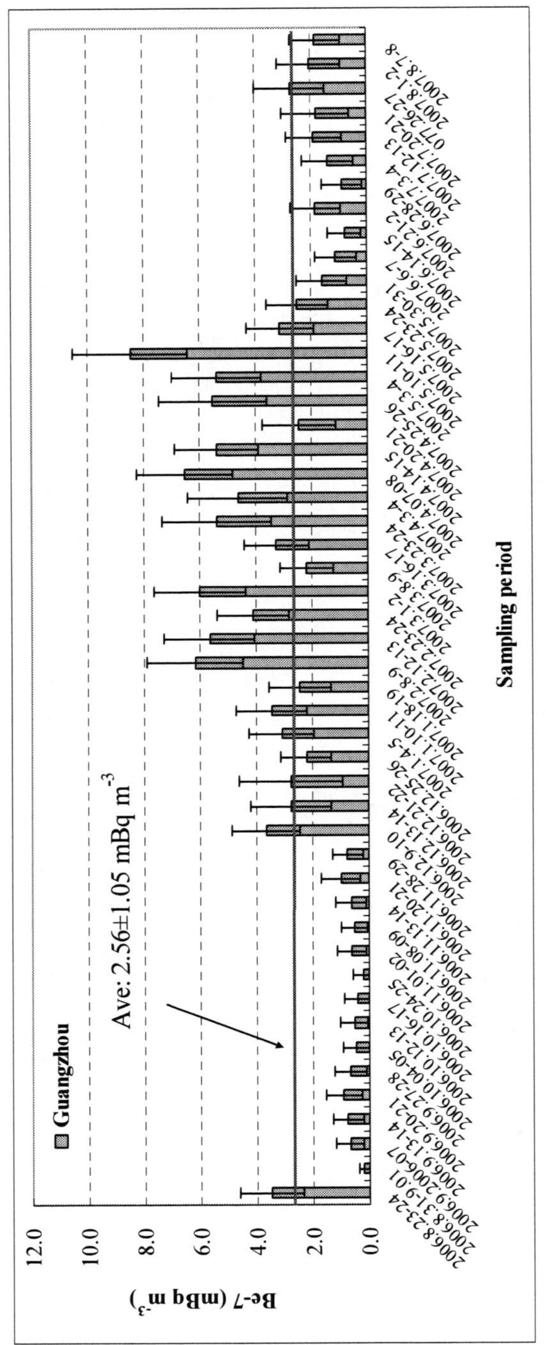

Figure 3. Comparisons of ^{7}Be concentrations in near-surface atmospheric aerosols in Beijing, Qingdao from August 2009 to July 2010, and Guangzhou from August 2006 to July 2007 on a weekly basis (the error bars denote 1σ) (Tan et al., 2013; Yang et al., 2013; Pan et al., 2011).

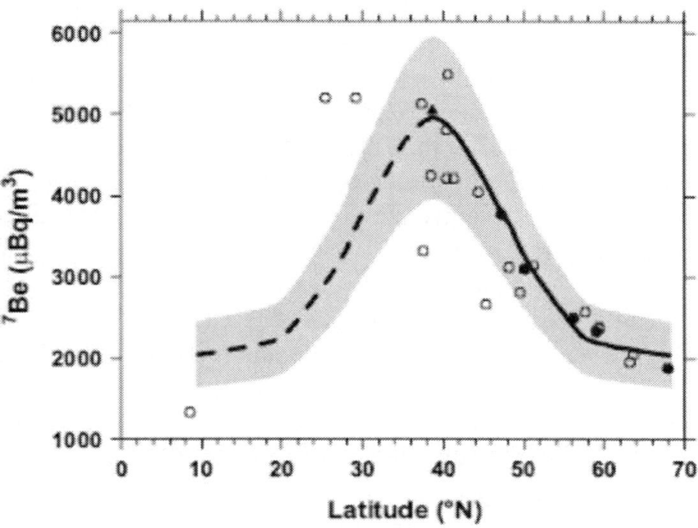

Figure 4. Average ^{7}Be activity of as a function of latitude (quoted from Kulan et al., 2006).

2009-2010 were a period of a particular minimum of solar activity, which could cause higher ^{7}Be in atmosphere than in ordinary years. Therefore it is better to compare with other cities in the East Asian monsoon climate system (EAMCS) and in the world during the same sampling period. Annual mean ^{7}Be concentrations in atmosphere in several other regions in the world during 2009-2010 have been reported. The annual mean concentration of ^{7}Be in atmospheric aerosols in Rio de Janeiro, Brazil (22°33'S) was 0.6 mBq m^{-3} (Pacini et al., 2011). In Thessaloniki, Northern Greece (40°38' N), 6.01 mBq m^{-3} was reported (Ioannidou, 2011). The monthly mean concentration of ≤ 7 mBq m^{-3} in Belgrade (44°49' N), Serbia were observed (Todorovic et al., 2005). Therefore it is evident that the latitudinal effect on near-surface level of ^{7}Be in atmospheric aerosols showed no exception during the low solar activity year within the same atmospheric circulation domain, but the annual mean concentrations of ^{7}Be in atmospheric aerosols in Beijing was relatively higher compared with those at similar latitude in non-EAMCS regions of the world.

Our result is generally in accordance with that of Kulan et al. (2006) and the fallout distribution of cosmogenic isotopes to the Earth's surface characterized by a mid-latitude peak (Bourcie et al., 2011; Pacini et al., 2011). The annual average ^{7}Be activity at the station of Guiyang appeared to be higher due to the high altitude of this station (1100 m above the sea level) (Wan et al., 2006).

Compared with the global curve (Kulan et al., 2006), the maximum of ^{7}Be in near-surface aerosols at the different latitude cities in China occurred at a slightly higher latitude (40°N) than that in the global curve (38.7°N), but the maximum value (8.39 mBq m^{-3}) is much higher than that (5.5 mBq m^{-3}) in the global curve. It should be pointed out that our annual means data did not cover the higher latitude (i.e., north of 40°N). However, from the following discussion on short-term simultaneous measurements of ^{7}Be in five cities in EAMCS, it can be seen that the data of Heihe city (50°N) may provide an evidence of lowered ^{7}Be concentration in atmospheric aerosols at higher latitude, suggesting that the mass movement in the East Asian monsoon zone is more convergent at mid-latitude due to the strong roles of Hadley Cell and Ferrel Cell (Fig. 1), causing distribution of atmospheric ^{7}Be more focusing at mid-latitude.

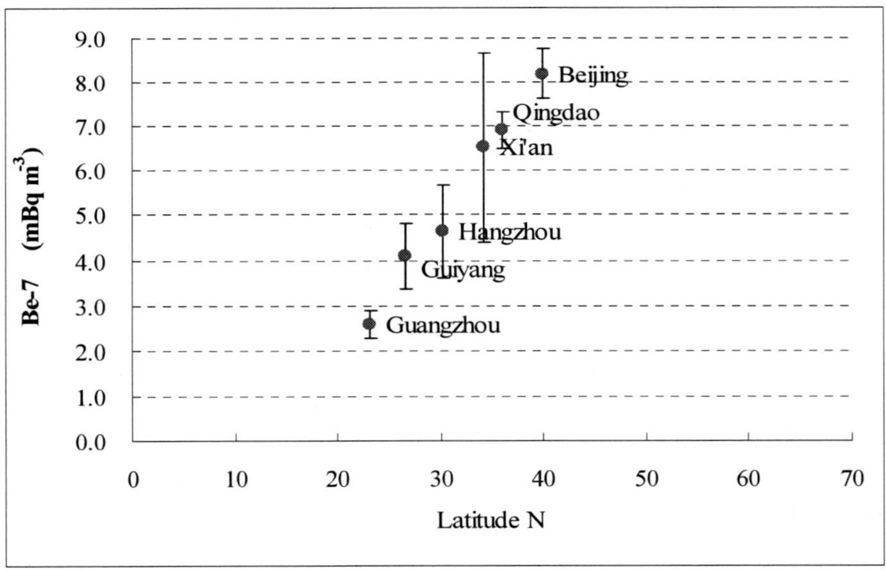

Figure 5. Latitudinal distribution of annual means of concentrations of ^{7}Be in near-surface atmospheric aerosols in East Asian monsoon zone in China (the error bars denote 1σ).

Table 2. Annual mean ^{7}Be concentrations in Chinese cities at different latitude.

City	Altitude (m) a.s.l.	Latitude (°N)	^{7}Be (mBq/m^3)	Error ($\pm1\sigma$)	Reference
Beijing	31.3	39.9	8.39	0.49	Tan et al., 2013
Qingdao	75	36.0	6.90	0.41	Yang et al., 2013
Xi'an	424	34.17	6.51	2.13	Chang *et al.*, 2008
Hangzhou	41.7	30.15	4.64	1.03	Jiang et al., 1994;
Guiyang	1100	26.57	4.09	0.72	Wan et al., 2006
Guangzhou	25	23.1	2.57	0.31	Pan et al., 2011

Simultaneous Measurement of ^{7}Be in Aerosols at Five Cities

The distribution in Fig. 5 is for annual average ^{7}Be only. A question arises how ^{7}Be in the atmosphere is distributed with latitude at a particular time in the East Asian monsoon zone? In order to better understand the latitudinal distribution of ^{7}Be in aerosols in the East Asian monsoon zone, we conducted simultaneous measurements of ^{7}Be in five cities in EAMCS. The results are presented in Table 3. Figure 6 presents the latitudinal distribution pattern of near-surface atmospheric ^{7}Be measured during spring (October 19-22, 2009), autumn (April 24-30 2010), summer (the Summer Solstice 2013), and winter (the winter Solstice 2013).

These latitudinal distributions of ^{7}Be showed that the distribution pattern still approximately followed a Gaussian curve-like pattern, but compared with the annual average ^{7}Be distribution pattern, the maximum concentrations were significantly higher in spring and autumn (15.2±0.9 mBq m^{-3} and 10.9±0.64 mBq m^{-3}, respectively), than in summer and winter, with the peaks occurring at Suzhou (31.2°N), i.e., the peak moving southward. From the air mass back-trajectory, it can be seen that a cold air mass moved from Siberian region during the autumn sampling period (Fig. 7). Air mass from Siberian region during the formation of the Siberian anticyclone can intensify the stratosphere-to-troposphere exchange (STE)(Lee et al., 2004) and strongly influence the weather in the East Asian monsoon zone during autumn, winter and early spring. The average ^{7}Be concentration of the Summer Solstice and the Winter Solstice in atmospheric aerosols in Suzhou was low (3.54 and 6.96 mBq m^{-3}, respectively), which well compensated the higher concentrations in spring and autumn periods in Suzhou.

Table 3. Simultaneous measurement of ^{7}Be in near-surface atmospheric aerosols at five cities.

City	Latitude °N	Sampling period	Mean ^{7}Be (mBq m^{-3})	Error ($\pm 1\sigma$) (mBq m^{-3})
Autumn 2009				
Guangzhou	23.1	2009/10/19-2009/10/23	4.82	0.27
Suzhou	31.2	2009/10/20-2009/10/23	7.23	0.14
Qingdao	36.04	2009/10/19-2009/10/24	10.00	0.61
Beijing	39.54	2009/10/19-2009/10/23	10.92	0.64
Heihe	50.22	2009.10.19-2009.10.24	8.88	0.52
Spring 2010				
Guangzhou	23.1	2010/4/25-2010/4/29	6.93	0.55
Suzhou	31.2	2010/4/25-2010/4/29	15.19	0.91
Qingdao	36.04	2010/4/24-2010/4/29	14.58	0.81
Beijing	39.54	2010/4/25-2010/4/30	8.41	0.47
Heihe	50.22	2010/4/25-2010/4/29	10.64	0.64
2012 Summer Solstice				
Guangzhou	23.1	2012/6/18-2012/6/25	3.38	0.38
Suzhou	31.2	2012/6/19-2012/6/25	3.54	0.37
Qingdao	36.04	2012/6/13-2012/6/25	4.37	0.35
Beijing	39.54	2012/6/19-2012/6/25	6.45	0.52
Heihe	50.22	2012/6/19-2012/6/24	3.92	0.59
2012 Winter Solstice				
Guangzhou	23.1	2012/12/17-2012/12/25	2.37	0.27
Suzhou	31.2	2012/12/20-2012/12/26	6.96	0.67
Qingdao	36.04	2012/12/18-2012/12/24	6.31	0.53
Beijing	39.54	2012/12/19-2012/12/25	6.87	0.57
Heihe	50.22	2012/12/18-2012/12/24	3.52	0.94

Factors Influencing the Seasonal Variation of ^{7}Be

In contrast, the maximum ^{7}Be concentrations were significantly lower on the Summer Solstice and the Winter Solstice 2013 (6.45±0.52 mBq m^{-3} and 6.87±0.57 mBq m^{-3}, respectively) compared with those in spring and autumn, with the peaks occurring at Beijing on the Summer Solstice and Suzhou on the Winter Solstice, i.e., the peak moving northward in summer and southward in winter. The high ^{7}Be concentration in spring may be related to the narrower tropopause. The stratopause becomes lower and the troposphere narrower in mid-latitude during spring months (from February to April) when the stratosphere-troposphere exchanges become strengthened and ^{7}Be in the stratosphere can be more easily leaked into the troposphere, i.e., "spring leak maximum" occurs (Andrews and Fontes, 1992; Stohl et al., 2000).

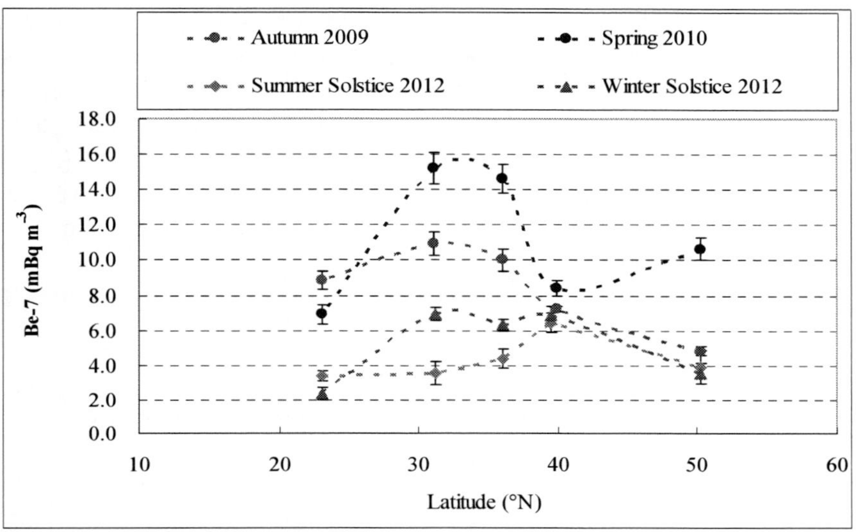

Figure 6. Short-term latitudinal distributions of ^{7}Be in aerosols in different Chinese cities in East Asian monsoon zone during autumn 2009, spring 2010, the Summer Solstice 2013, and the Winter Solstice 2013 (the error bars denote 1σ).

According to the atmospheric circulation theory, the air mass near the Equator receives more heating than air masses at other latitude, and forming an up-rise movement. The air mass moved toward higher latitude while moving to the high elevation in the troposphere (Fig. 1). Geostrophic force will produce a westward velocity which becomes progressively greater towards higher latitude and the component of velocity toward the polar direction will become smaller, leading to a horizontal convergence and sinking of the air masses, i.e., the sinking branch of the Hadley Cell. The air mass at high elevation moving towards lower latitude regions in the Ferrel Cell sinks at subtropical zone due to the air continuum principle, forming the sinking branch of the Ferrel Cell (Zhang et al., 2010). Although the Hadley Cell and the Ferrel Cell exist in all seasons, the cells can move southward during winter and northward in summer, in accordance with the solar effective radiation strength (Zeng et al., 2011).

Northwesterlies are predominant in winter and spring in EAMCS. The ^{7}Be concentrations in winter were relatively constant, reflecting a stable origin of air mass. The air masses from high latitude regions to EAMCS during November-April period may have origins in Mongolia and Siberia. Based on ^{7}Be concentrations in atmospheric aerosols of different particle size in Hong

Kong, Lee et al. (2004) found that [7]Be source was restricted in regions of Mongolia and the southeastern Siberia, which supported the proposal that the [7]Be source area was associated with the anti-cyclone in Siberia. In contrast, marine-modified air masses originated from the West Pacific Ocean in late June and early July when the intrusion of western Pacific High occurs (Fig. 2). Besides, during the summer monsoon season, four channels of strong cross-equatorial flows located within 40°E–135°E are found to bring clean air to China from the Southern Hemisphere (Zhang et al., 2010). Air masses from low latitude regions through these channels have the effect of diluting [7]Be concentrations in the eastern China.

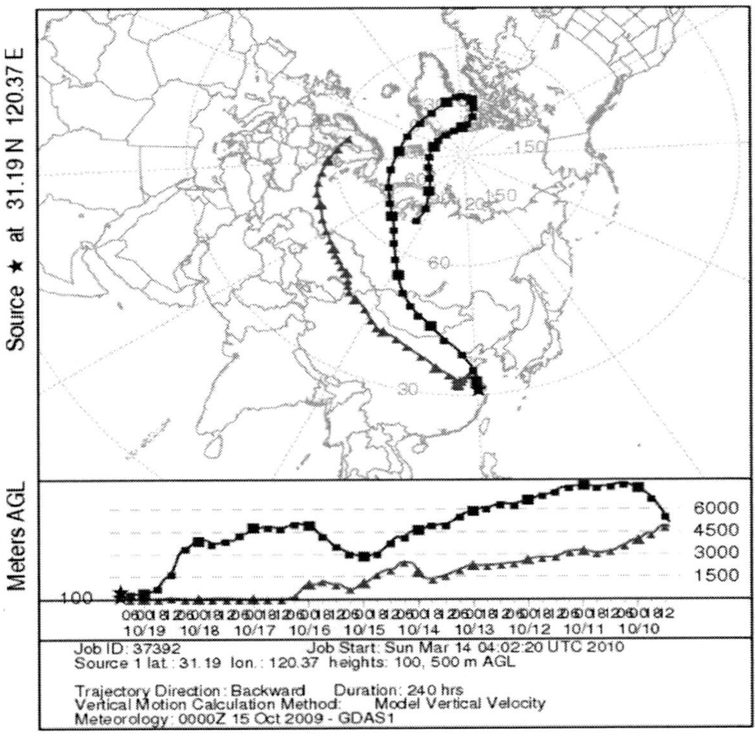

Figure 7. Back-trajectories of air-masses at the sampling sites in Suzhou during the sampling period. (from http:// www.noaa.arl.gov, USA) (Colors denote different air-mass movement and the figures in the lower frames represent altitude in meter). Five-day back trajectories were calculated using the Hybrid Single-Particle Lagrangian Integrated Trajectory (HYSPLIT) model.

In autumn 2009, the global tropical and sub-tropical atmospheric circulation showed significant adjustment, causing abnormal climate globally, with the sub-tropical high weaker in the high latitude regions. Affected by the El Nino events occurring from June 2009 to spring 2010 continuously in the mid- and eastern equatorial Pacific, beginning from July 2009, the West Pacific Subtropical High was remarkably stronger both in area and strength than those in the same periods in ordinary years. From August 2009, the West Pacific Subtropical High obviously invaded westwards (Zeng et al., 2011). These events could promote air exchange between the lower stratosphere and the troposphere, causing increase of atmospheric ^{7}Be in the troposphere, leading to higher ^{7}Be concentrations in near-surface atmospheric aerosols in autumn 2009.

Implications on Latitudinal Distributions of OCPs and PCBs in Aerosols in EAMCS

Fig. 8(a) presents the concentrations of various groups of POPs studied in the same aerosol samples of ^{7}Be simultaneously collected at the five stations of different latitude on 19-24, October 2009. OCPs and PCBs had been officially banned in use and production since 1980s in China, therefore, the OCPs and PCBs found in atmospheric aerosols should be mostly evaporated from the residues in soils or long-range transported from other regions.

PCBs were the dominant compounds in these POPs in aerosols at all stations (51.0-194.9 pg m^{-3}). Higher PCBs were found in Suzhou and Beijing, with the maximum occurring in Suzhou and the minimum in Qingdao. HCHs were observed highest in Suzhou, and DDTs showed a maximum in Beijing. The three groups of POP compounds were all low in Qingdao that is located in the Shandong Peninsular north of the Yellow Sea. Influenced by the oceanic climate, air masses in Qingdao come from the ocean generally containing less aerosols, playing a dilution effect to the aerosols in continental air mass. Lammel et al. (2007) investigated POPs in atmospheric aerosols in the Yellow Sea in summer and found that DDTs, HCB, HCHs, and PCBs concentrations were in the range of 16–180 pg m^{-3}. Our results for DDTs, HCHs, and PCBs in atmospheric aerosols in Qingdao during the sampling period were 10–200 pg m^{-3}, therefore the low concentrations of these POPs in aerosols in Qingdao could not be an occasional phenomenon.

Yong-Liang Yang, Xiao-Hua Zhu, Nan Gai et al.

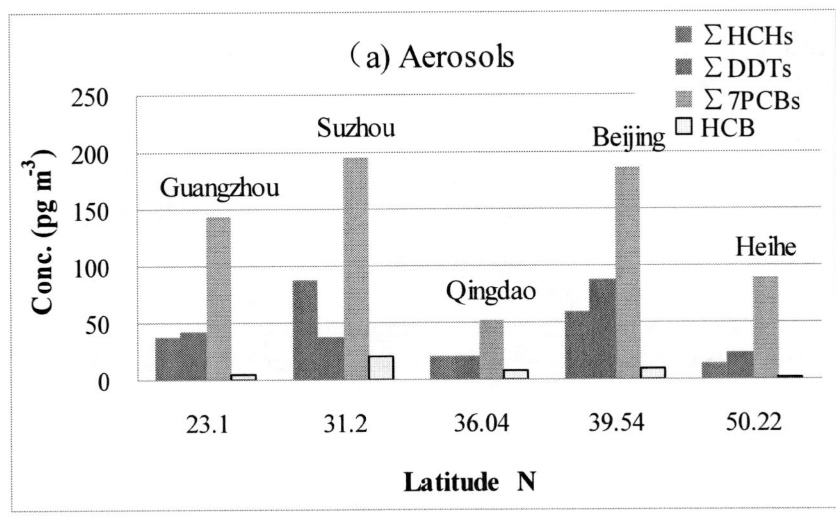

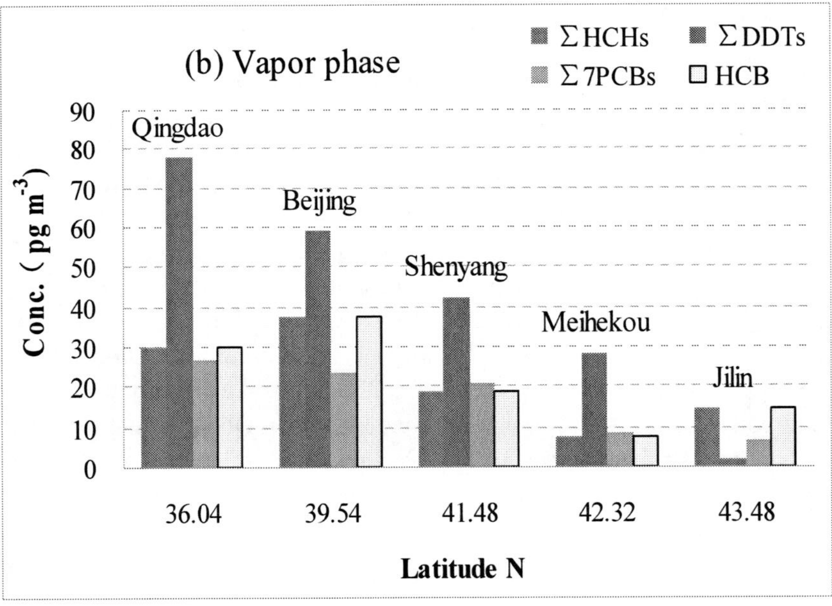

Figure 8. Latitudinal distributions of concentrations of POP compounds in (a) aerosols and (b) vapor phase from cities of different latitude in East Asian monsoon zone in Autumn, 2009.

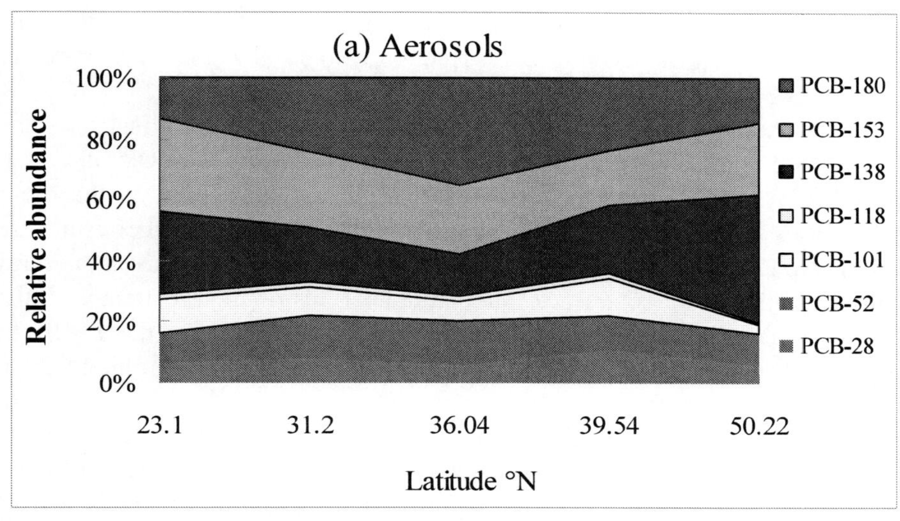

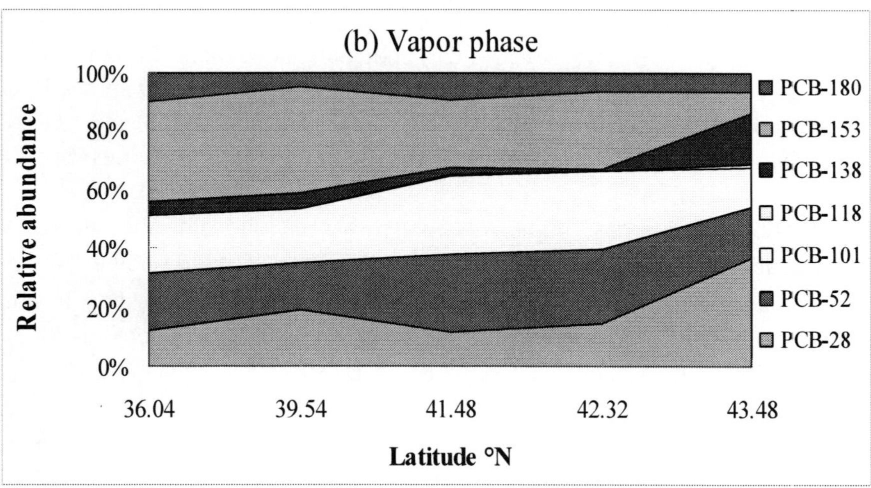

Figure 9. Latitudinal distributions of compositions of PCBs in (a) aerosols and (b) vapor phase from cities with different latitude in East Asian monsoon zone in autumn, 2009.

With the exception for the lower POPs concentrations in aerosols in Qingdao, the highest concentration of various POP compounds in aerosols in the East Asian monsoon zone in autumn of 2009 occurred at the mid-latitude stations Beijing and Suzhou, especially HCHs and PCBs had the maximum occurring in Suzhou, similar to the case of [7]Be. It has been reported that the distribution pattern of PCBs in surface soils approximately shows a Gaussian distribution globally with 45°N as the center whereas HCB in surface soils approximately shows a Gaussian distribution with 60°N as the center (Breivik et al., 2002).

As discussed above, the concentrations of POPs may be affected by the dilution of relatively cleaner air mass. In order to compare the POPs' levels in aerosols with those in vapor phase, passive air samplings were conducted in five cities (Qingdao, Beijing, Shenyang, Meihekou, Jilin) and the results are shown in Fig. 8(b). DDTs were the dominant compounds in passive samplings at all stations. In contrast to the latitudinal distribution patterns of the concentrations in aerosols, the latitudinal distribution patterns of these POPs in vapor phase showed a distinctive trend, i.e.; the concentrations of the POPs in general decrease with increase of latitude.

If relative abundance of POP compounds instead of their concentrations is used, some information on the behaviors of POP compounds with different volatility could be revealed. The latitudinal distributions of abundance of individual PCB congeners in aerosols and vapor phase at the five stations are shown in Fig. 9. In the seven PCB indicator congeners in atmospheric aerosols, the higher chlorinated PCB138, PCB153, and PCB180 showed high relative abundances in lower latitude city Guangzhou. PCB153 and PCB180 were dominant PCB congeners in aerosols in mid-latitude cities Suzhou, Qingdao, and Beijing.

PCB28 which has higher vapor pressure, showed the maximum concentration in vapor phase at the station of highest latitude (Heihe, 50° N), in accordance with the phenomenon that PCB28 is easier to transport to polar areas (Wania, 2003). In comparison, the variation in PCB28 in aerosols did not change with latitude significantly, while PCB180 in aerosols which has low tendency of evaporation, is basically accumulated around 36° N, showing some focused accumulation effect. On the other hand, the concentrations of PCB180 in vapor phase basically did not change with latitude. PCB138 in aerosols had the highest abundance at 50°N. Our results are basically in consistent with the classification of POPs proposed by Wania (2003). The vapor pressures of HCHs and PCBs are relatively higher than those of DDTs with higher long-range transport potentials (Wania, 2003). Based on the chemical partitioning space defined by the two equilibrium partitioning coefficients K_{AW} and K_{OA}, PCB28 can be regarded as a flier (substances that are too volatile to partition appreciably into the surface compartments water and soil) and PCB180 and p,p'-DDT as single hoppers (substances that are too less volatile to re-evaporate after they have been deposited to the Earth's surface)(Wania, 2003).

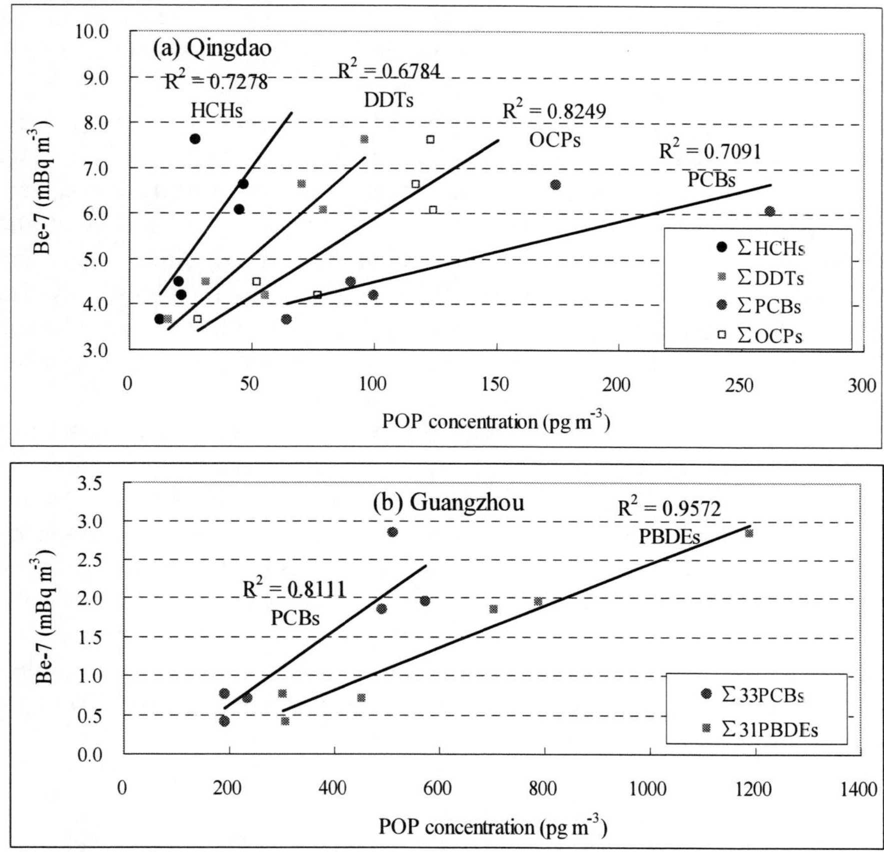

Figure 10. Correlations between the monthly average concentrations of ^{7}Be with (a) $\sum$HCHs, $\sum$DDTs, $\sum$PCBs, and $\sum$OCPs during January, February, March, May, June, and July, 2010 in near-surface atmospheric aerosols of Qingdao, and (b) $\sum_{31}$PCBs, and $\sum_{33}$PBDEs in near-surface atmospheric aerosols of Guangzhou during the period of July ~ December, 2007 (Pan et al., 2011; Yang et al., 2013).

Correlations between POP Compounds and ^{7}Be in Atmospheric Aerosols

During the sampling period of September 2009, January, February, March, May, June, and July 2010 in Qingdao, ΣHCHs, ΣDDTs, and ΣPCBs showed good co-relationship with ^{7}Be (Fig. 10a) (Yang et al., 2013), suggesting that these POP compounds in atmospheric aerosols in Qingdao may have similar transport mode with ^{7}Be in these months, leading to higher consistency in their seasonal variation patterns. Similar correlations had been also found in Guangzhou during the period of July ~ December, 2007 (Fig. 10b)(Pan et al., 2011). However, these POP compounds in atmospheric aerosols in Qingdao showed no co-relationship with ^{7}Be in April when a sharp increase of ^{7}Be occurred due to the strengthened stratosphere-troposphere exchanges, i.e.; "spring leak maximum" (Todorovi et al., 2005; Yong and Silker, 1980; Andrews and Fontes, 1992; Stohl et al., 2000), and in August and September 2009, when the West Pacific Subtropical High obviously invaded westwards (Shen and Chen, 2011) and several typhoon events occurred, which could promote air exchange between the lower stratosphere and troposphere, causing increase of atmospheric ^{7}Be in troposphere.

In summary, the above discussions that the relative abundance of low volatile POP compounds in atmospheric aerosols co-vary to a certain extent with the latitude variations of ^{7}Be in the atmosphere aerosols. As the distributions of ^{7}Be in the atmosphere are the functions of atmospheric circulations and meteorological conditions, independent of human activities, it can serve as an effective tracer for pollutants in atmospheric circulations and as a reference framework to compare with the latitudinal distributions of POPs, so as to avoid the perplexity due to irregularity of geographical distribution of unknown sources of pollutants in predicting the long-term geographic distribution of POPs in soils and biota.

Implications for Latitudinal Distributions of POPs in Surface Soils in EAMCS

Figure 11 shows the latitudinal distributions of PCBs, HCHs, and DDTs in surface soils in East Asian monsoon zone in China found from literature (numbers of data for PCBs and OCPs are n=40 and n=27, respectively, some data represents the average concentration of several sampling sites in one area, i.e., some data are the means for areas). The striking features of the latitudinal distributions of PCBs, HCHs, and DDTs in surface soils are the two peaks for the compounds, with the maximum concentrations occurring at 30°N and 40°N (OCPs) and 44°N (PCBs). The zone around 30°N is the most populated regions such as the Yangtze River Delta area. The zone around 40°N is important agriculture and industrial region represented by Beijing-Tianjin-Tangshan industrial area and the Northeast China's industrial area. These zones may have experienced maximum usage of these compounds. However, the concentrations of HCHs, DDTs, and PCBs in the other important regions for agriculture and industries such as the Pearl River Delta region (~ 20°N) where is also one of the most populated region in China were low. Also the concentrations of HCHs and DDTs in China's most important agricultural production region---Heilongjiang Province (north of 45°N) were also low, where annual temperature is the lowest in China and the SOM (soil organic matter) in soil are high which may help preserve POPs in soils (Meijer et al., 2003). Latitudinal distributions of concentrations of POP compounds in passive air samplings from cities of different latitude in East Asian monsoon zone in autumn (Fig. 8b) provide evidence that the population and the historical usage of OCPs and PCBs are not the only factors controlling these chemicals' latitudinal distribution. Therefore, the atmospheric circulations in EAMCS are likely to have been influential in causing this observation. On the other hand, the latitudinal distribution mode of POPs in atmosphere and surface soils may be different in different climate system.

CONCLUSION

The results for investigation of ^{7}Be in near-surface atmospheric aerosols in the East Asian monsoon zone in China reveal that the latitudinal distribution of the annual average ^{7}Be concentrations follows a normal distribution pattern, with the maximum occurring at ~40°N. The instantaneous observation of ^{7}Be at different latitude sampling sites in EAMCS also showed a normal distribution pattern, but with the maximum at 30°N in winter.

The latitudinal distribution pattern can instantaneously move southward or northward at a particular time depending on season. The value at the peak will also vary seasonally. The short-term latitudinal distributions of near-surface atmospheric ^{7}Be display the peaks at latitudinal range of 30° - 40°N, depending on the strength of the sinking air from the high elevation and the extents to which the air exchange between troposphere and stratosphere.

Spring and autumn were seasons with high near-surface atmospheric ^{7}Be in the East Asian monsoon zone and the peaks were at 30°N due to the strong wind direction changes in seasonal transition periods. On the other hand, near-surface atmospheric ^{7}Be concentrations were lower in winter and summer while the peak position shifted northward at 40°N.

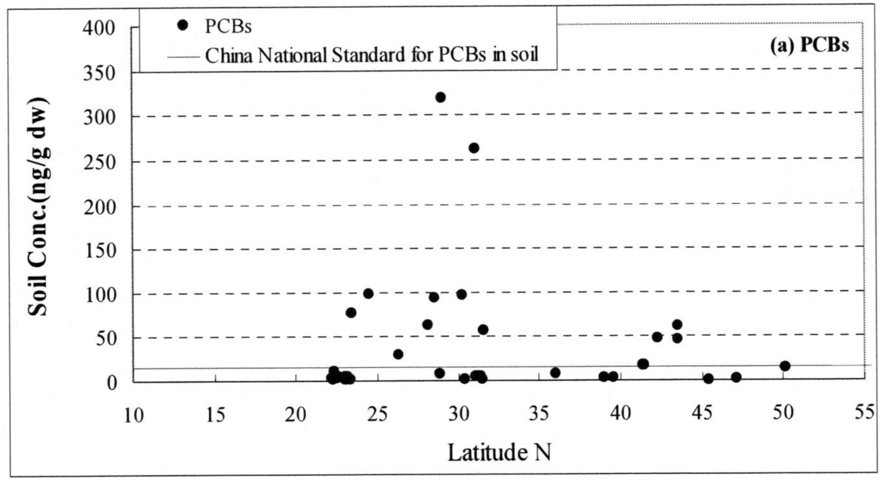

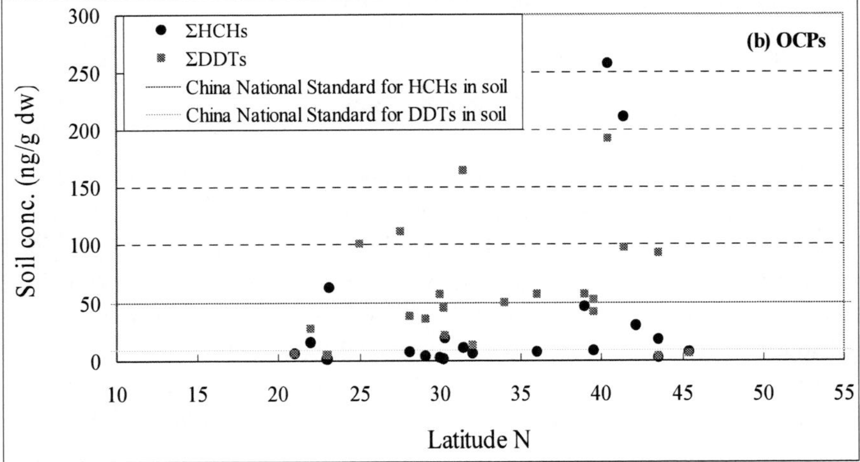

Figure 11. Latitudinal distributions of (a) OCPs and (b) PCBs in surface soils in East Asian monsoon zone in China. The solid lines denote the thresholds for concentrations of PCBs, HCHs, DDTs by Chinese National Standard for Soil Quality. (Numbers of areas where the data were averaged for OCPs and PCBs are n=27 and n=40, respectively).

The maximum concentrations of OCPs and PCBs in near-surface aerosols occurred at different latitude showed a similar pattern as that of ^{7}Be, i.e., with the maximum at 30~40°N. The abundance of PCB28 in aerosols showed the maximum at 40°N. PCB180, which has low evaporation tendency, in aerosols showed enrichment at 36°N, while PCB180 in vapor phase appeared to be nearly unchanged with latitude, suggesting focused accumulation mechanisms for POP compounds with different volatile tendencies in the atmosphere in EAMCS. Our results suggest that the atmospheric circulations are likely to have been influential in causing this observation and the latitudinal distribution mode of POPs in atmospheric aerosols in EAMCS. Our result may provide reference for latitudinal distribution of air pollutants associated with aerosols such as radioactive nuclide fallout emitted due to nuclear power station accident to the Earth's surface in the East Asian monsoon zone. Further research is needed with higher resolution both in temporal and spatial scales.

ACKNOWLEDGMENTS

We thank Suzhou Environmental Monitoring Station and Guangzhou Institute of Geochemistry, Chinese Academy of Sciences, for help atmospheric aerosol samplings in Suzhou and Guangzhou. This work was supported by the Natural Science Foundation of China (Grant 41073011, 410773010, and 41003044)

REFERENCES

Aguado, E., Burt, J.E. 2010. *Understanding Weather and Climate*, Prentice Hall, London, 608 pp.

Akata, N., Kawabata, H., Hasegawa, H., Sato, T., Chikuchi, Y., Kondo, K., Hisamatsu, S., Inaba, J. 2008. Total deposition velocities and scavenging ratios of ^{7}Be and ^{210}Pb at Rokkasho, Japan, *J. Radioanal. Nucl. Chem.,* 277, 347–355.

Ali, N., Khan, E. U., Khan, F., Akhtar, P., Waheed, A. 2010. Determination of aerosol mean residence time using ^{210}Pb and ^{7}Be radionuclides in the atmosphere of Islamabad. *The Nucleus*, 47, 25–29.

Andrews, J. N., Fontes, J. C. 1992. Importance of in-situ production of chlorine-36, argon-36 and carbon-14 in hydrology and hydrogeochemistry. In: Isotope Techniques in Water Resources 1991, SM-319. International Atomic Energy Agency, Vienna, 245pp.

Ayub, J. J., Di Gregorio, D. E., Velasco, H., Huck, H., Rizzotto, M., Lohaiza, F. 2009. Short-term seasonal variability in ^{7}Be wet deposition in a semiarid ecosystem of central Argentina. *J. Environ. Radioact.*, 100, 977–981.

Bleichrodt, J.F. 1978. Mean tropospheric residence time of cosmic ray produced beryllium-7 at north temperate latitude, *J. Geophys. Res.*, 83, 3058–3062

Bourcier, L., Masson, O., Laj, P., Pichon, J. M., Paulat, P., Freney, E., Sellegri, K. 2011. Comparative trends and seasonal variation of ^{7}Be, ^{210}Pb and ^{137}Cs at two altitude sites in the central part of France, *J. Environ. Radioact.*, 102, 294–301.

Breivik, K., Sweetman, A., Pacyna, J. M., Jones, K. C. 2002. Towards a global historical emission inventory for selected PCB congeners-a mass balance approach: 1. Global production and consumption. *Sci. Total. Environ.*, 290, 199–224.

Chang, Y.Z., Wang, X. H., Wang, S. L., Wang, J. 2008. Radionuclides monitoring in atmospheric aerosol samples in Xi'an, *Nucl. Tech.*, 31, 796–800 (In Chinese with English abstract).

Dutkiewicz, V.A., Husain, L. 1985. Stratospheric and tropospheric components of ^{7}Be in surface air, *J. Geophys. Res.*, 90, 5783–5788.

Elliot, S., Blake, D. R., Duce, R. A., Lai, C. A., McCreary, I., McNair, L. A., Rowland, F. S., Russell, A. G., Streit, G. E., Turco, R. P. 1997. Motorization of China implies changes in pacific air chemistry and primary production. *Geophys. Res. Lett.*, 24, 2671-2674.

Feely, H. W., Larsen, R. J., Sanderson, C. G. 1989. Factors that cause seasonal variations in Beryllium-7 concentrations in surface air, *J. Environ. Radioact.*, 9, 223–249.

Garimella, S., Koshy, K., Singh, S. 2003. Concentration of ^{7}Be in surface air at Suva, Fiji. South Pacific. *J. Natural Sci.*, 21, 15–19.

Gouin, T., Mackay, D., Jones, K. C., Harner, T., Meijer, S.N. 2004. Evidence for the "grasshopper" effect and fractionation during long-range atmospheric transport of organic contaminants. *Environ. Pollut.*, 128, 139–148.

Hansen, J. C. 2000. Environmental Contaminants and Human Health in the Arctic. *Toxicol. Lett.*, 112/113, 119–125.

Hasegawa, H., Akata, N., Kawabata, H., Chikuchi, Y., Sato, T., Kondo, K., Inaba, J. 2007. Mechanism of ^{7}Be scavenging from the atmosphere through precipitation in relation to seasonal variations in Rokkasho Village, Aomori Prefecture, Japan, *J. Radioanal. Nucl. Chem.*, 273, 171–175.

Holton, J. R., Haynes, P. H., McIntyre, M. E., Douglass, A. R., Rood, R. B., Pfister, L. 1995. Stratosphere -troposphere exchange, *Rev. Geophys.*, 33, 403–439.

Igarashi, Y., Hirose, K., Otsujihatori, M. 1998. Beryllium-7 deposition and its relation to sulfate deposition. *J. Atmos. Chem.*, 29, 217–231.

Ioannidou, A. 2011. Activity size distribution of ^{7}Be in association with trace metals in the urban area of the city of Thessaloniki, Greece, *Atmos Environ.*, 45, 1286–1290.

Jiang, R.R. 1994. The ^{7}Be concentration in the air above ground, *Nucl. Tech.*, 17, 172–175 (In Chinese with English abstract).

Koch, D. M., Jacob, D. J., Graustein, W. C. 1996. Vertical transport of tropospheric aerosol as indicated by ^{7}Be and ^{210}Pb in a chemical tracer model. *J. Geophys. Res.*, 101, 18651–18666.

Kulan, A., Aldahan, A., Possnert, G., Vintersved, I. 2006. Distribution of ^{7}Be in surface air of Europe, *Atmos. Environ.*, 40, 3855–3868.

Lal, D., Malthotra, P. K., Peters, B. 1958. On the production of radioisotopes in the atmosphere by cosmic radiation and their application to meteorology, *J. Atmos. Terres. Phys.*, 12, 306–328.

Lamborg, C. H., Fitzgerald, W. F., Graustein, W. C., Turekian, K. K. 2000. An examination of the atmospheric chemistry of mercury using ^{210}Pb and ^{7}Be. *J. Atmos. Chem.*, 36, 325–338.

Lammel, G., Ghim, Y.S., Gradosa, A., Gao, H., Huhnerfuss, H., Lohmann R. 2007. Levels of persistent organic pollutants in air in China and over the Yellow Sea. *Atmos. Environ.*, 41, 452–464.

Lee, L.Y.L., Kwok, R. C. W., Cheung, Y. P., Yu K. N. 2004. Analyses of airborne ^{7}Be concentrations in Hong Kong using back-trajectories, *Atmos. Environ.*, 38, 7033–7040.

Li, Y., Geng, X. C., Yu, H. Q., Wan, G. J. 2011. Effects of the composition of standard reference material on the accuracy of determinations of ^{210}Pb and ^{137}Cs in soils with gamma spectrometry, *Appl. Radiat. Isotopes*, 69, 516–520.

Lin, W.L., Zhu, T., Tang, X. Y. 2003. Observation of atmospheric tracer Be-7 in the Mt. Qomolangma Region, *Acta Scientiarum Naturalium Universitatis Pekinensis*, 39, 51–55 (in Chinese with English abstract).

Meijer, S. N., Ockenden, W. A., Sweetman, A., Breivik, K., Grimalt, J. O., Jones, K. C. 2003. Global distribution and budget of PCBs and HCB in background surface soils: Implications for sources and environmental processes. Environ. Sci. Technol., 37 (4), 667-672.

Momoshima, N, Nishio, S. Kusano, Y., Fukuda, A., Ishimoto, A. 2006. Seasonal variations of atmospheric ^{210}Pb and ^{7}Be concentrations at Kumamoto, Japan and their removal from the atmosphere as wet and dry depositions, *J. Radioanal. Nucl. Chem.*, 268, 297–304.

Ockenden, W. A., Sweetman, A. J., Prest, H. F., Steinnes, E., Jones, K. C. 1998. Towards an understanding of the global atmospheric distribution of persistent organic pollutants: the use of semi-permeable membrane devices as time-integrated passive samplers. *Environ. Sci. Technol.*, 32, 2795–2803.

Ockenden, W. A., Breivik, K., Meijer, S. N., Steinnes, E., Sweetman, A. J., Jones K. C. 2003. The global re-cycling of persistent organic pollutants is strongly retarded by soils. *Environ. Pollut.*, 121, 75–80.

Pan, J., Yang, Y. L., Zhang, G., Shi, J. L., Zhu, X. H., Li, Y., Yu, H. Q. 2011. Simultaneous observation of seasonal variations of beryllium-7 and typical POPs in near-surface atmospheric aerosols in Guangzhou, China, *Atmos. Environ.*, 45, 3371–3380.

Pacini, A. A., Usoskin, I. G., Evangelista, H., Echer, E., de Paula, R. 2011. Cosmogenic isotope ^{7}Be: A case study of depositional processes in Rio de Janeiro in 2008–2009, *Adv. Space Res.*, 48, 8114–818.

Shen, H. Y. Chen, Y. B. 2011. Analysis of origin of drought in upreach of the Yangtze River during 2009~2010. *Yangtze River*, 42, 12–14.

Shi, Z. L., Wen, A. B., Yan, D. C., Zhang, X. B., Li, J. 2011. Temporal variation of ^{7}Be fallout and its inventory in purple soil in the Three Gorges Reservoir region, China, *J. Radioanal. Nucl. Chem.*, 288, 6715–676.

Stohl, A., Spichtinger-Rakowsky, N., Bonasoni, P., Feldmann, H., Memmesheimer, M., Scheel, H. E., Trickl, T.. Hubener, S., Ringer, W., Mandl, M. 2000. The influence of stratospheric intrusions on alpine ozone concentrations, *Atmos. Environ.*, 34, 1323–1354.

Tan, K. Y., Yang, Y. L., Zhu, X. H., Chen, S., Jiao, X. C., Gai, N., Huang, Y. 2013. Beryllium-7 in near-surface atmospheric aerosols in mid-latitude (40 degrees N) city Beijing, China,. *J Radioanal. Nuc. Chem.*, 298, 883–891.

Todorovic, D., Popovic, D., Djuric, G., Radenkovic, M. 2005. Be-7 to Pb-210 concentration ratio in ground level air in Belgrade area, *J. Environ. Radioact.*, 79, 297–307.

Todorovic, D., Popovic, D., Nikolic, J., Ajtic, J. 2010. Radioactivity monitoring in ground level air in Belgrade urban area, *Radiat. Protec. Dosimetry,* 142, 308–313.

UNEP. 2001. Final act of the plenipotentiaries on the Stockholm Convention on persistent organic pollutants. *United Nations environment program chemicals.* Geneva, 445 pp.

Van Aardenne, J. A., Carmichael, G. R., Levy II, H., Streets, D., Hordijk, L. 1999. Anthropogenic NOx emissions in Asia in the period 1990-2020. *Atmos. Environ.,* 33, 633–646.

Wan, G. J., Zheng, X. D., Lee, H. N., Wang, S. L., Wan, E. Y., Yang, W., Tang, J., Wu, F. C. C., Wang, S., Huang, R. R. 2006. A comparative study on seasonal variation of ^{7}Be concentrations in surface air between Mt. Waliguan and Mt. Guanfeng, *Geochimica,* 35, 221–226.

Wania, F. 2003. Assessing the potential of persistent organic chemicals for long-range transport and accumulation in polar regions. *Environ. Sci. Technol.,* 37, 1344–1351.

Wania, F., Mackay, D. 1993. Global fractionation and cold condensation of low volatility organochlorine compounds in polar regions. *Ambio,* 22, 10–18.

Wania, F., Mackay, D. 1996. Tracking the distribution of persistent organic pollutants. *Environ. Sci. Technol.,* 30, 390A–396A.

Wania, F., Shen, L., Lei, Y. D., Teixeira, C., Muir, D. C. G. 2003. Development and calibration of a resin-based passive sampling system for monitoring persistent organic pollutants in the atmosphere. *Environ. Sci. Technol.,* 37, 1352–1359.

Yang, Y.L., Gai, N., Geng, C. Z., Zhu, X. H., Li, Y., Xue, Y., Yu, H. Q., Tan, K. Y. 2013. East Asia monsoon's influence on seasonal changes of beryllium-7 and typical POPs in near-surface atmospheric aerosols in mid-latitude city Qingdao, China, *Atmos. Environ.,* 79, 802–810.

Yong, J.A., Silker, W. B. 1980. Aerosol deposition velocities on the Pacific and Atlantic oceans calculated from ^{7}Be measurements, *Earth Planet. Sci. Lett.,* 50, 92–104.

Yoshimori, M. 2005. Production and behavior of beryllium-7 radionuclide in the upper atmosphere. *Advances in Space Res.,* 36, 922–926.

Zeng, G., Wang, W. C., Sun, Z. B., Li, Z. X. 2011. Atmospheric circulation cells associated with anomalous East Asian winter monsoon. *Advances Atmos. Sci.,* 28, 913–926.

Zhang, L., Liao, H., Li, J. P. 2010. Impacts of Asian summer monsoon on seasonal and interannual variations of aerosols over eastern China, *J. Geophys. Res.,* 115, D00K05, doi:10.1029/2009JD012299.

Zheng, X. D., Wan, G. J., Tang, J., Zhang, X. C., Yang, W., Lee, H. N., Wang, C. S. 2005. ^{7}Be and ^{210}Pb radioactivity and implications on sources of surface ozone at Mt. Waliguan, *Chinese Science Bulletin 50*, 1675–171.

In: Aerosols
Editor: Adam Yanick Pearson

ISBN: 978-1-63117-512-1
© 2014 Nova Science Publishers, Inc.

Chapter 3

AEROSOLS PROCESSING OF NANOSTRUCTURED OXIDES FOR ENVIRONMENTAL APPLICATIONS

M. Miki-Yoshida[*]*, P. Amézaga-Madrid,*
W. Antúnez-Flores, P. Pizá-Ruiz, C. Leyva-Porras,
C. Ornelas-Gutiérrez and O. Solís-Canto
Centro de Investigación en Materiales Avanzados S. C.,
Laboratorio Nacional de Nanotecnología, Chihuahua, México

ABSTRACT

Nanostructured materials and coatings have gained great importance due to their microstructural properties and their use has increased in different technological areas such as: environmental pollution control, photocatalysis, optics, solid oxide fuel cells, electronic and optoelectronic devices, mechanical protection, catalysis, biomedical applications, etc. Furthermore, advances in aerosol processing in recent years have increased the variety of produced nanostructured materials, including metals, oxides, ceramics, composites, and fullerenes; in different appearances such as nano-particles, rods, belts, fibers, tubes, needles, films, etc. This chapter presents the synthesis of nanostructured oxides in the form of nanoparticles, and multilayered or composite thin films by

---

[*] Corresponding author: M. Miki-Yoshida. Tel.: +52 614 439 1114, Fax: +52 614 439 4824, E-mail address: mario.miki@cimav.edu.mx.

aerosol assisted chemical vapor deposition technique. It is well established that atomic arrangement and the microstructure of materials determine their properties. Interestingly, similar morphologies cut across all material classes and the different methods used to synthesize them; thus, controlling composition, grain size, morphology, and crystallinity are primary concerns of the synthesis and processing of materials.

In consequence, a microstructural characterization of nanostructured materials and coatings is crucial in the optimization of processing in order to obtain a desired property or behavior. Therefore, a detailed microstructural characterization of these materials was performed by electron microscopy and x-ray diffraction. In the case of films and multilayers, surface micro-structure was analyzed by field emission scanning electron microscopy (FESEM); also, cross-sectional microstructure were also analyzed by FESEM and high resolution transmission electron microscopy (HRTEM). Cross sectional samples for HRTEM analysis were prepared by focused ion-beam technique. Elemental composition of the materials was determined by energy dispersive x-ray spectroscopy with corresponding systems attached to electron microscopes. Crystalline structure was analyzed by grazing incidence x-ray diffraction.

The optical characterization of the thin films was attained by UV-VIS-NIR spectroscopy; optical constants and thickness were also determined. The optical properties were discussed in relation to the microstructural characteristics of the samples. Some of these materials were utilized for environmental pollution control applications. Films were employed for liquid phase photocatalysis of model pollutants; and nanostructured powders were used for the removal of As from aqueous solutions.

Keywords: Aerosol processing, Nanostructured materials, Electron Microscopy, Environmental pollution control

1. INTRODUCTION

Advances in aerosol processing in recent years have increased the variety of produced nanostructured materials, including metals, oxides, ceramics, composites, and fullerenes; and also in different presentations such as nanoparticles, rods, belts, fibers, tubes, needles, films, etc. In particular, nanostructured materials and coatings have gained great importance due to their interesting microstructural properties and their use has increased in different technological areas such as: environmental pollution control, photocatalysis, optics, solid oxide fuel cells, electronic and optoelectronic devices, mechanical protection, catalysis, biomedicine, etc.

Predominantly, thin films and coatings become more and more relevant in many disciplines, such as: optics, solid oxide fuel cells, electronic and optoelectronic devices, mechanical protection, catalysis, etc.

Several applications depend of processes occurring at or near the surface; it is then convenient to use reduced volumes of the active material, generally expensive, deposited as a thin layer onto adequate support, mechanically strong and inexpensive. Thin films and coatings have a variety of applications in different areas such as solar energy exploitation and energy saving. In this field, one of the most common applications rely on transparent conductors (TC) [1] for solar control or low emittance windows, making use of the low infrared transmittance of the coatings. The materials for these applications are thin films based on metals, or on a wide band gap doped oxide semiconductors, such as In_2O_3, SnO_2 or ZnO. TCs can show relatively high transparency in the $0.4 < \lambda < 0.7$ μm (visible light) wavelength interval. In the infrared (IR) $\lambda > 0.7$ μm their metallic property leads to reflectance, and in the ultraviolet (UV) $\lambda < 0.4$ μm, they become absorbent due to excitations across an energy gap. Other applications are related with solar cells and "smart windows", in which the electrical conductivity of the transparent conductor films is employed to collect electrical charge in the appliance. Related applications deal with photocatalytic (TiO_2, ZnO), hydrophilic, thermochromic (VO_2) and electrochromic (WO_3) films and coatings. Another new and interesting application of oxides concerns the diluting of ferromagnetic oxides of particular interest for spintronic devices (or spin-based electronics) such as spin valves [2, 3].

These new magnetic materials are a group of diluted ferromagnetic oxides, which are generally n-type, high dielectric constant, wide-gap semiconductors with Curie temperatures T_C well in excess of room temperature; they have a very promising combination of material properties, such as semiconducting behaviour, transparency and ferromagnetism. Interest in these materials was triggered by a prediction of room temperature ferromagnetism in Mn doped ZnO, in the work of Dietl et al. [4]. After that, the first experimental reports for oxides were for Co-doped thin films of TiO_2 (anatase), ZnO and SnO_2 [5]. Spintronic devices aim to combine both the charges, as in conventional semiconductor devices, and the spins of electrons, as in magnetic materials. Here, the spin of electrons can be used as an added degree of freedom in novel electronic devices. Interestingly for oxides, this magnetic behaviour has only been observed in thin films and nanostructures. One of the latest new materials recently "discovered" is a naturally occurring two-dimensional carbon film known as graphene. It is indeed the thinnest free standing film ever identified [6]. Graphene is a single or a few layers of sp^2-hybridized carbon atoms, densely packed into a benzene-ring structure; with adequate morphological adjustments, it is the essential component of many carbon-based materials, including graphite, large fullerenes, nanotubes, etc. Graphene was discovered almost a decade ago and proved the existence of a two-dimensional system, which becomes sChart as a result of 3D corrugation, since it has been presumed not to exist in the free sChart state. It appeared that this material had exceptional electronic, mechanical, thermal and optical properties, for a broad range of applications. Since then, a lot of effort has been devoted to the study of graphene and to its fabrication [7]. In the mechanical protection area, thermal barrier coatings (TBCs) such as 8 wt.% Y_2O_3–ZrO_2 (Y8SZ) are able to protect the blades of a gas turbine engine and can lead to further increases in operating temperature by providing an effective thermal barrier on the blades [8]. In addition, dual gas-chamber micro-solid oxide fuel cells (μ-SOFC) with a planar geometry are promising power sources for porChart electronic devices. The μ-SOFC membranes are developed on the basis of thin film deposition and micropatterning [9]. Solid state chemical sensors were obtained with SnO_2 and In_2O_3 thin films, structural engineering of these metal oxides were performed for the optimization of gas sensing characteristics [10].

This chapter presents the synthesis of nanostructured oxides in the form of nanoparticles, multilayered or composite thin films by aerosol assisted chemical vapor deposition technique. A detailed microstructural characterization of these materials was performed by electron microscopy and x-ray diffraction. Surface morphology and microstructural characteristics were analysed by field emission scanning electron microscopy (FESEM). Cross sectional microstructure of the films and multilayers were also analysed by FESEM and by high resolution transmission electron microscopy (HRTEM). In general, the grain size and structural morphology were determined in order to correlate with structure-zone models reported in the documents. The elemental composition of the films was determined by energy dispersive x-ray spectroscopy with corresponding systems attached to electron microscopes. The inter-diffusion of components between film-substrate or interlayers was analysed by HRTEM of the cross sections of the samples, in this case, samples were prepared by focused ion beam technique. The crystalline structure was analysed by grazing incidence x-ray diffraction (GIXRD). The influence of deposition parameters on materials characteristics was theoretically and experimentally analysed. The optical characterization of the materials was realized by UV-VIS-NIR; optical constants and thickness were determined. The optical properties were discussed in relation to the material microstructure. These materials were utilized for some environmental pollution control applications. The films were employed for gas and liquid phase photocatalysis of model pollutants; and nanostructured powders were used for the removal of As from aqueous solutions.

2. AEROSOL PROCESSING OF NANOSTRUCTURED MATERIALS

The synthesis of thin films and nanostructured materials is the crucial step for the creation of devices and assemblies for many important applications, since many of them are all based on material structures composed of thin films. For example, the semiconductor industry is totally dependent on the formation of thin solid films of a variety of materials by deposition from the gas, vapour, liquid, or solid phase. Many techniques and methods exist for the synthesis of thin films and coatings, in the thickness range up to about few µm. Traditionally, they have been classified into physical and chemical methods; however, considerable number of techniques combine both physical processes and chemical reactions; these can be categorized as physical-chemical methods [11]. Other classification schemes consider the scale at which the material is deposited, independently of the mechanism or processes involved [12]. In this representation, four categories are proposed: atomistic growth, particulate deposition, bulk coating, and surface modification.

The physical methods denoted in general as PVD (physical vapour deposition) describe a variety of methods to deposit thin solid films by the condensation of a vaporized form of the solid material onto various surfaces. PVD involves physical ejection of material as atoms or molecules, its condensation and nucleation onto a substrate; eventually ejected material reacts with gases to form new compounds. For chemical methods, chemical vapour deposition (CVD) represents perhaps the most relevant and extended technique. In contrast to physical methods, it is a process whereby a solid material is deposited from a vapour caused by a chemical reaction occurring in the vicinity of a normally heated substrate surface. There are several CVD modifications, utilizing different mechanisms for the transport of the reactant vapour to the substrate or procedures to enhance or activate the chemical reaction.

In this work, we will consider the aerosol processing CVD variant where the process performs at atmospheric pressure, with the substrate maintained at a relatively high temperature and where the reactant, in the form of an aerosol, is carried towards the substrate by a gas. This method has several denominations in the pertaining documents, such as aerosol assisted CVD (AACVD), atmospheric pressure CVD, spray pyrolysis, etc. Table 1 resumes the most important deposition techniques. Rigorously, there is not a sharp division between different methods.

Aerosol assisted chemical vapour deposition is a well-known method for the synthesis of thin films and powders [13]. It is a variant of conventional CVD technique, the most important differences are: a) generally AACVD is performed at atmospheric pressure; b) the transport of precursors is attained by the aid of an aerosol flow towards the reaction zone; i.e., the precursor solution is converted in an aerosol and conveyed by a gas carrier. The advantages of this method in comparison to conventional CVD are the relative simplicity and low cost of implementing the system, easy dopant incorporation, can obtain many types of materials such as composites, multi-layered, core-shell structures and, on the other hand, the scalability to industrial processes.

Table 1. Classification of deposition methods according to processes involved in the synthesis

Physical	Chemical
Sputtering: magnetron, RF	Chemical vapour deposition (CVD) atmospheric-pressure, low-pressure, metalorganic
Evaporative	Electrophoretic deposition
Cathodic arc deposition	Electroplating
Molecular-beam epitaxy	Electroless plating
Electron-beam evaporation	
Pulsed laser deposition	

Physical- Chemical
Reactive: sputtering, evaporation, cathodic arc.
CVD variants: aerosol processing, plasma enhanced, photo enhanced, electron enhanced.
Electrostatic spray-assisted vapour deposition
Sol Gel: dip and spin coating

2.1. Synthesis of Thin Films by Aerosol Assisted CVD

Aerosol assisted chemical vapour deposition [14] is a CVD variant in which the reactants are transported as a mist of the precursor solution, conveyed and directed towards the heated substrate. At the substrate's surface, the gaseous precursor decomposes and reacts to generate the nucleation and growth of the film. The precursor solutions used are a dilution of organometallic or inorganic compounds in a solvent, such as alcohol or distilled water. The mist of the precursor solution (aerosol) was generated by a particular type of nebulizer (ultrasonic, electrostatic, or pneumatic). The aerosol is conveyed by a carrier gas towards the substrate by a nozzle, the flow of the carrier gas was optimized for each material.

For single components, there is a broad substrate temperature interval where the desired materials are obtained with uniformity, adherence, smoothness, homogeneous microstructure and properties. In contrast, for binary or ternary compounds, this temperature interval is reduced to a more specific optimum temperature [15], dependent on the thermodynamic properties of the precursor compounds.

The nozzle has a periodic movement at a constant velocity to scan the whole surface of the substrate to get uniform deposits. The thickness was controlled by scan velocity and a number of steps. If the film is deposited in several steps, certain materials are deposited as a multi-layered structure, as it is described in the microstructural analysis section. Some materials need a thermal treatment after deposition for complete crystallization and stabilization of their microstructure [15].

This technique was also used to deposit nanostructured films of ZnO and TiO_2 on the internal surface of fused silica tubing [16].

Solid Works – Fluid Works were used to perform simulation of the aerosol flow towards the substrate considering the distribution of the temperature around the heating plate, particularly to assess the actual temperature of the aerosol stream and flow velocity near to the surface; parameters which are determinant in the uniformity and growth rate of the film. Aerosol temperature and velocity were considered the same as the carrier gas, due to the small size of the aerosol droplets (2 μm).

Results of the distribution of aerosol (fluid) temperature and component (normal to the substrate) of aerosol velocity are presented in Figure 1. It is shown that heating of the aerosol takes place very near to the substrate surface; in addition the cooling effect of the carrier gas flux can be noticed.

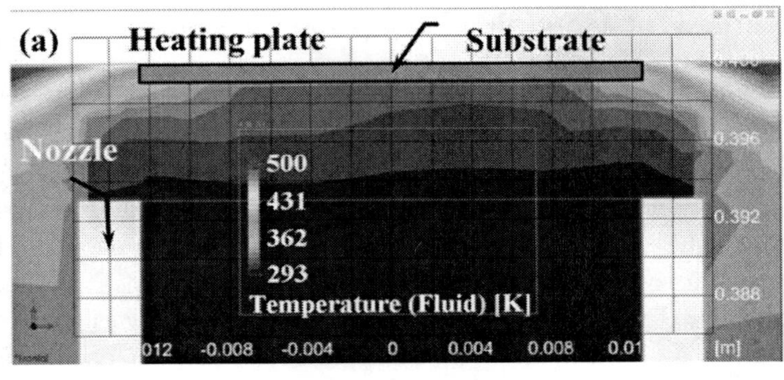

a

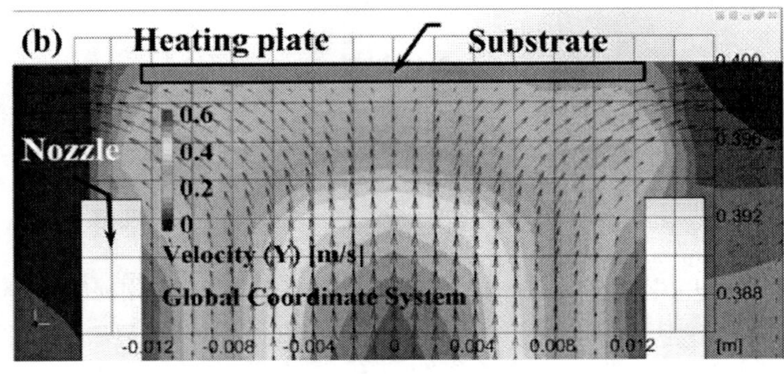

b

Figure 1. Simulation results of aerosol temperature a) and normal component of its velocity b) around the substrate surface. Arrows depict the direction and magnitude of total velocity. Other experimental conditions were: heating plate temperature 523 K; carrier gas flow of 8 L/min; distance nozzle-substrate 6 mm.

Our films were obtained using a pneumatic (C, H, I, J, K) or an ultrasonic nebulizer Sonaer model PG-241 working at 2.4 MHz. The carrier gas was micro-filtered air at a fixed pressure of 125 kPa; and flow rates between 4 - 15 L.min⁻¹. Samples C and H were deposited on the internal surface of fused silica tubing.

In this chapter, we have analysed several metal oxide thin films obtained onto borosilicate glass and fused silica substrates (2.5×2.5 cm^2) by AACVD technique. Table 2 presents the principal preparation parameters of the films analysed in this chapter.

Table 2. Synthesis parameters of the films analysed. Films C, H, I, J, and K were obtained using a pneumatic nebulizer; the other with an ultrasonic nebulizer Sonaer model PG-241 working at 2.4 MHz. Samples C and H were deposited on the internal surface of fused silica tubing. Carrier gas was micro-filtered air at a pressure of 125 kPa

Sample	Material at. ratio D/M	Precursors	Substrate temperature [K]	Carrier gas flow [L.min^{-1}]	Ref.
A	ZrO$_2$:Y at. Y/Zr= 0 – 0.25	Zr acetylacetonate Y acetate	723	5	[14, 17]
B	PbTiO$_3$ at. Pb/Ti=1	Ti oxyacetylacetonate Pb acetate	623 - 673	4	[15]
C	ZnO	Zn acetate	725	4	[16]
D	ZnO:Co at. Co/Zn = 0 – 0.1	Zn acetate Co acetate	623 - 723	5	[18]
E	ZnO:Sc at. Sc/Zn = 0 – 0.24	Zn acetate Sc acetylacetonate	523 - 723	5	[19]
F	ZnO:V at. V/Zn = 0 – 0.13	Zn acetate V acetylacetonate	523 - 723	5	[19]
G	ZnO:Mg at. Mg/Zn = 0 – 1	Zn acetate Mg acetate	673 - 723	5	-
H	TiO$_2$	Titanyl acetylacetonate	775	4	[16]
I	TiO$_2$:Cu Cu/Ti = 0.04 – 0.15	Titanyl acetylacetonate Cu (II) acetate	673	4	[20]
J	TiO$_2$:Al at. Al/Ti = 0.05 – 0.35	Titanyl acetylacetonate Al acetylacetonate	723	8	[20]
K	SnO$_2$ 0.05 – 0.5	Tin (IV) chloride	523 - 673	15	[21]

2.2. Synthesis of Nanostructured Powders by Aerosol Assisted CVD

Metal oxide nanostructured powders were obtained inside of fused silica tubing by AACVD technique, details of the experimental set up have been described previously [22]. The method is analogous to that used to deposit thin films; the funda-mental difference is that, for powder synthesis, the aerosol is injected by the carrier gas directly in the reaction chamber (tubing), where decomposition and reaction of precursors take place to produce the desired nanostructured material. The collection of the powders is performed in the outermost side of fused silica tubing into an electrostatic precipitator or a liquid collector.

Solid Works – Fluid Works were used to perform simulation of the aerosol flow towards the reactor tubing considering the distribution of the temperature inside the tubular furnace, to determine the temperature distribution of the aerosol stream and flow velocity. Aerosol temperature and velocity were considered the same as the carrier gas, due to the small size of the aerosol droplets (2 μm). Representative results of the distribution of aerosol (fluid) temperature and aerosol velocity are presented in Figure 2. The effect of the carrier gas flow on the heating of the aerosol has been verified; i.e., its maximum temperature shifts to the exit as the flow rate increases. Simulated temperature distributions were compared with experimental measurements.

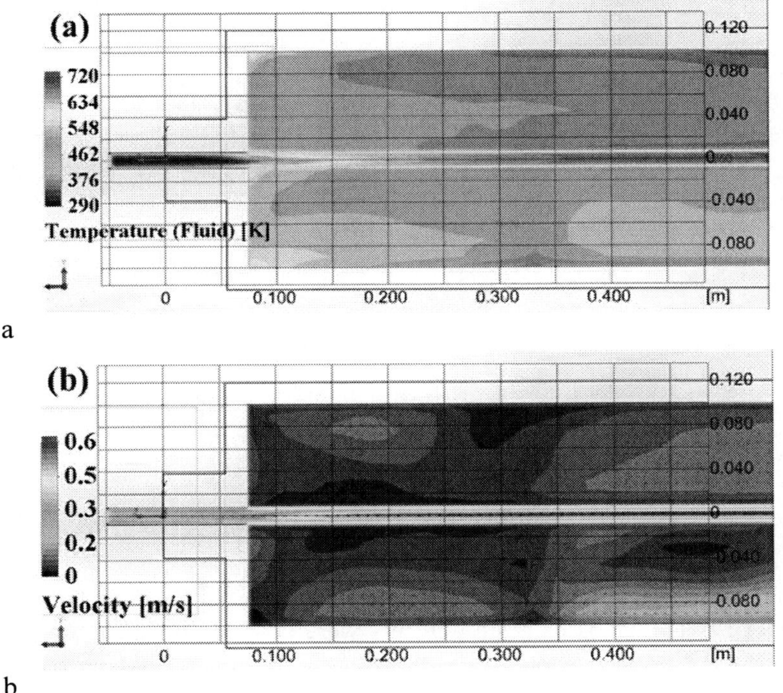

Figure 2. Simulation results of temperature distribution a) and gas velocity b) inside the tubular furnace and heated reactor tubing.

Table 3 presents the deposition parameters of the nanostructured powders analysed in this chapter.

Table 3. Synthesis parameters of the nanostructured powders analysed

Sample	Material	Precursors Concentration [mol.dm^{-3}]	Substrate temperature [K]	Carrier gas flow [L.min^{-1}]	Ref.
L	Fe_3O_4	Fe (II) chloride 0.01 – 0.1	723 – 873	Ar: 1 Air: 0.015	[22]
M	$Ca_{0.14}Zr_{0.86}O_2$	Zr acetylacetonate Ca acetate	1073 - 1273	2 - 20	[23]

3. MICROSTRUCTURAL ANALYSIS OF THIN FILMS AND NANOSTRUCTURED OXIDES OBTAINED BY AEROSOL ASSISTED CVD

Science and technology have advanced to a stage in which microscopic characterization is no longer enough to appropriately elucidate a phenomenon or process; now it is necessary to push forward into the nanoworld, and use nanoscience and nanotechnology to develop and manufacture innovations needed in daily life.

It is well established that atomic arrangement and the microstructure of materials determine their properties; thus controlling composition, grain size, morphology, and crystallinity are of primary concern in the synthesis and processing of materials, particularly of thin films and coatings. In consequence, a detailed microstructural characterization of thin films and coatings is crucial in the optimization of processing in order to obtain a desired property or behaviour.

Electron microscopy (EM) in every one of its varieties constitutes one of the most powerful techniques for the microstructural characterization and analysis of virtually any type of solid materials (like metals, semiconductors, ceramics, polymers, composites) in every conformation, for instance bulk, powder, thin film, etc. Recently, developments of new detectors make possible the analysis of wet samples at low vacuum, i.e., for biological samples. These techniques can give information about morphology, composition, crystalline structure, and electronic structure; at very high resolution, up to atomic level.

Several types of electron microscopes exist, such as scanning, transmission and dedicated scanning transmission electron microscopes. Concurrently they give complementary information of the microstructural properties of the analysed material. Other techniques profited from the methods of electron microscopy and incorporated them; for example focused ion beam (FIB) incorporates electron detectors to obtain images of the samples' working zones. Thus, electron microscopy renders possible the nexus between process of synthesis, microstructural evolution and properties of the materials. In this sense, electron microscopy techniques assist in the control of microstructural characteristics, like composition, grain size, morphology and crystallinity of the material with the objective to optimize their properties.

3.1. Experimental Methods for Microstructural Characterization

Microstructural characteristics of the films, such as composition, crystalline phases, thickness, grain morphology, crystallite size, interfacial features, among others, were obtained as a function of several deposition parameters. Analogously, nanostructured powders were analysed to determine composition, crystalline phases, grain morphology, porosity, crystallite size and distribution, interfacial features, etc.

The surface morphology and cross sectional microstructure of the films were studied by field emission scanning electron microscopy (SEM) using a JEOL JSM-7401F, operated at 5 kV. From this analysis High angle annular dark field (HAADF) images in scanning transmission electron microscopy (STEM) mode were done in a JEOL JEM 2200FS with STEM Cs-corrector, operated at 200 kV. Selected cross sectional samples were also studied by high resolution transmission electron microscopy (HRTEM). SEM cross section samples were prepared by conventional methods in a cross section polisher JEOL SM-09010 at 5 kV and 120 µA. In addition, cross sectional samples for HRTEM analysis were prepared using focused ion beam (FIB) in a JEOL JSM 9320 system, with Ga ion beam operated at 30 kV. Cross sections were lifted out in situ by an Omniprobe 200 nanomanipulator. Elemental analysis of the films was achieved by means of energy dispersive X-ray spectroscopy (EDS), using an Oxford Inca microanalysis system attached to the electron microscopes. For planar surface analysis, some films were processed by ion milling etching (IME) to analyse different stages of the film growth in a Gatan 691 PIPS. For these studies the films were peeled off from the glass substrate by immersion in HF (40%), then they were floated, rinsed in deionized water, and were finally mounted on a TEM copper grid.

Grazing incidence x-ray diffraction patterns were used to determine the crystalline phases present on the films and nanostructured powders, using a Panalytical X-Pert system, working with Cu Kα radiation (λ = 0.1542 nm) at 40 keV and 30 mA. Grazing incidence angle was set at 0.5°, scanning angle 2θ was varied between 15° - 95°, with 0.1°/min of scan time, and a step size of 0.02°. Planar or cross sections of the films can be studied depending on the sample preparation technique used. Planar section of the sample was processed by ion milling etching, whereas cross sectional sample was obtained by focused ion beam method.

3.2. Growth and Microstructural Characteristics of Thin Films Deposited by Aerosol Assisted CVD. Cross Sectional Analysis of the Film. Composition and Elemental Distribution. Crystalline Structure of Thin Films. Crystallite Size

In this section, the growth characteristics and crystalline structure of characteristic thin films obtained by aerosol assisted CVD method is explained. Several growth, geometrical and crystallographic attributes of thin films and coatings were analysed by SEM, TEM, GIXRD and related techniques (EDS). Cross sectional analysis of typical thin films obtained by aerosol assisted CVD method gives straightforward information related with thickness, overall structure of single or multilayers coating, granular or columnar structure, also qualitatively about density or porosity, and roughness of the surface or interfaces. In particular, the observation of columns has been used in recent years to create thin films with three-dimensional nanostructures, such as chevron-type or helical ones, with unusual optical, electrical, magnetic, mechanical properties [24]. In addition, qualitative information about surface and interface roughness can be obtained and compared to other measurements, for example by scanning probe microscopy. Cross sectional analysis of films and coatings can be performed in a SEM or TEM system. SEM analysis has the advantage of less sample preparation labour; however, obviously resolution is lower (~1 nm). For better results, samples have to be prepared by a cross section polisher to obtain a flat section. In the case of HRTEM, sample preparation is a delicate and tedious procedure, since it is not ease to get ready a 100 nm thick sample without any microstructural modification. Several established techniques are employed for TEM sample preparation, including ion milling etching and focused ion beam.

It is well established that the deposition rate in a CVD process presents typical regimes as a function of temperature [25]. For deposition temperatures lower than a certain critical value $T_s < T_c$ there is a rather linear relationship between the deposition rate (r) and the inversion of the temperature ($1/T_s$). This is a typical Arrhenius behavior of an activated process, in which the deposition rate is limited by the reaction kinetics process at or near the substrate surface. i.e., chemisorption, and/or chemical reaction, surface migration, lattice incorporation and desorption.

$$r = K \cdot e^{-\frac{E_a}{RT_s}} \tag{1}$$

Where K is independent of T_s, E_a is the apparent activation energy, R is the gas constant and T_s is the deposition temperature [25]. Other region corresponds to $T_s > T_c$, where the deposition rate is almost constant and reaches its highest value. It is the typical process in which the growth rate is mass transport controlled, and the deposition rate depends weakly on temperature. The reaction kinetics is so fast that the surface reaction finally becomes controlled by the mass transfer of the reactants, i.e., diffusion of the active gaseous species through a boundary layer to the substrate surface. At even high temperatures ($T_s \gg T_c$), the deposition rate may decrease due to the depletion of reactants and/or increase in the rate of desorption. Other possible reason may be due to the occurrence of an alternative reaction involving high temperature etching generated by corrosive reactants or by-products.

Furthermore, at high temperatures homogeneous gas phase reactions can occur in addition to the heterogeneous reaction, decreasing the overall deposition efficiency of the film. This will lead to particle formation in the gas phase, and to the deposition of poorly adherent films with a non-uniform and/ or porous microstructure. This homogeneous gas phase reaction constitutes the principle for the production of nanostructured powder materials. In general, films deposited in the reaction kinetic limit are smoother than those deposited in the mass transport regime.

Thus, depending of the application of the film, these two regimes can be used to synthesize smooth or rough surfaces. For example, in photocatalytic processes it should be more convenient for rough surfaces to enhance the specific surface area of the film.

In the particular case of AACVD, the aerosol size and size distribution plays a crucial role in the quality of the deposited films. For the deposition of a determined material, i.e., fixed precursor solution nature (solute and solvent), concentration, aerosol generation method, and aerosol characteristics; it can be possible to find the other specific optimized deposition conditions, such as T_s, carrier gas flow, geometrical parameters, to obtain high quality films [26]. This means that the solvent has to be evaporated and the solute precipitate that forms afterward undergoes volatilization near the substrate surface. Then, adsorption of the vapor onto the heated substrate surface, followed by its decomposition and/or chemical reactions, yields the deposition of the desired film. This mechanism is similar to the heterogeneous CVD deposition process, which produces dense films with excellent adhesion. However, if the aerosol size distribution is too wide, the optimum conditions for small droplets will be very different to those for larger ones. Consequently it is impossible to optimize the deposition conditions of too differently sized droplets in order to obtain good films. Therefore narrow size distribution of aerosols is a prerequisite for high quality materials; it is more important than the size of the aerosol. Narrow size distribution is one of the most important advantages of ultrasonic nebulization compared to the pneumatic method.

An important variable that influences the characteristics of a thin film is its thickness. Therefore, it is essential to compare films of approximately the same thickness to asses a certain property or to analyze the effect of a certain parameter of synthesis. The influence of a film's thickness can be justified, for example, in observations of SnO_2 film microstructure by transmission electron microscopy. This analysis indicates that the films start to grow in an amorphous manner; then, growth continues with the nucleation of many small crystallites (3 – 14 nm) and finally some of them grow until a closely packed micro-structure of crystallites is formed [21]. In the intermediate layer, grain sizes range between 25 and 120 nm; these grains were formed by crystallites with many defects, twin boundaries and dislocations. In the top layer, the grain size ranges between 60 and 170 nm; the grain boundaries are thin, with less defects. It also can be observed in many small clusters (3 – 7 nm), which should be nucleation centers, in the faces of some crystals. To analyze the different stages of the film growth, different ion milling etching processes were used: top IME for observing the bottom layers formed near the substrate, top-bottom IME for observing intermediate layers, and bottom IME for observing top layers of the films [21]. Due to the small thickness of the films, less than 500 nm, different IME were performed very carefully and for short period of time (a few minutes).

The first group of samples corresponds to yttrium stabilized zirconium oxide (YSZ) thin films, sample A in Table 2. Starting with the cross sectional analysis of a multi-layered YSZ. Figure 3a presents bright field STEM micrographs of the overall cross section of a multi-layered YSZ film; it is shown Pt-C and Cr layers deposited to protect the sample's surface, during FIB cross sectional preparation. From this analysis, thickness of the film or layer can be readily obtained with high precision, considering adequate scale calibration of the microscope. In this sample, total thickness of the coating was around 160 ± 8 nm; while low density layer (dark) had around 15 ± 3 nm and high density (bright) 5 ± 1 nm [14]. The coating consisted of a periodic stack of dense and porous YSZ layers of the same composition, as it was determined by line scan energy dispersive x-ray spectroscopy and nanobeam diffraction of individual layer [27]. A proposed explanation of the origin of this periodic microstructure was that the dense layers were produced locally where the nozzle injected the aerosol directly onto the substrate; whereas the mist of aerosol that surrounded the surface deposits gradually the porous layer while the nozzle traversed along the substrate [14]. In figure 3b is shown the atomic resolution of well crystallized grains observed in all the film, including porous and dense layers. AFM analysis of the film's surface determined that grain size was around 35 ± 6 nm, and RMS roughness 3.7 ± 0.4 nm [14].

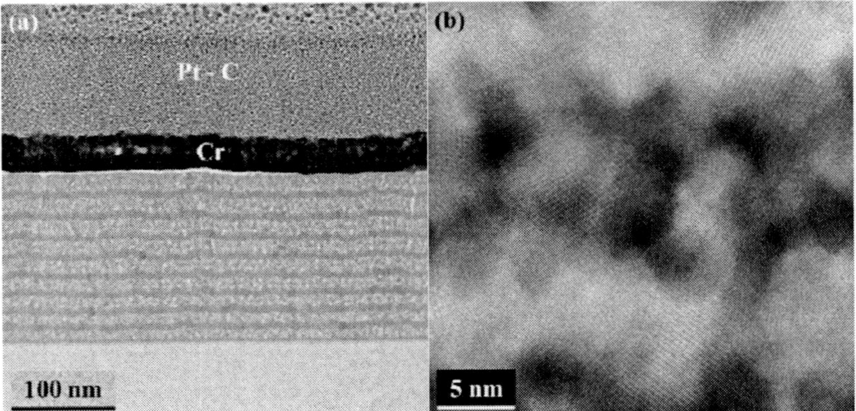

Figure 3. Bright field STEM images of the cross section of a multi-layered YSZ film. a) Overall structure of the film showing YSZ layers of different density (sample A). b) High resolution Z contrast image of low density (dark) and high density (bright) YSZ layers.

Growth rate of YSZ films prepared by AACVD was analyzed to determine the influence of aerosol velocity towards the substrate surface. It is known that growth rate is influenced by several parameters, such as: precursor concentration in starting solution, substrate temperature, droplet velocity towards the substrate (determined by the carrier gas flux, section of nozzle, distance between tip nozzle and substrate), lateral velocity of the nozzle v_n, and those related with the aerosol generation (such as size and size distribution of droplets).

$$t_{es} = \frac{d}{v_n} \qquad\qquad t_e = N \cdot t_{es} \qquad\qquad (2)$$

The growth rate of the films has been calculated considering an effective deposition time where the nozzle was injecting aerosol directly to a certain point on the surface, which gave an effective growth rate [14]. The effective deposition time t_e was estimated considering the number of steps N in the process, the lateral displacement velocity of the nozzle v_n, the section of the nozzle tip, and the distance between the nozzle tip and surface (see Figure 4).

A linear correlation ($R^2 \sim 0.98$) was observed between the effective growth rate and aerosol velocity; it can be stated that the effective growth rate increased linearly with the droplet velocity up to around 200 cm s^{-1}.

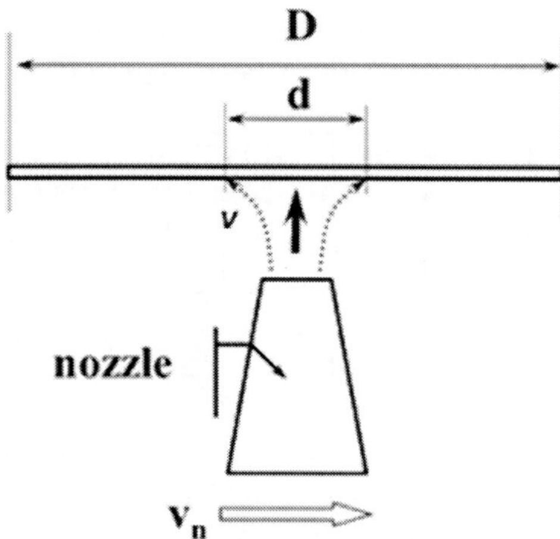

Figure 4. Parameters to define the effective growth rate. D=sample length, d=size of aerosol flux on the surface, v=aerosol velocity, v_n=lateral velocity of nozzle, t_{es}=effective time per step, N=number of deposition steps, t_e=effective deposition time.

The correlation between effective growth rate and droplet velocity can be justified considering that an increase of the aerosol normal velocity toward the substrate results in a more effective diffusion and adsorption of the precursor spices. Table 4 presents experimental droplet velocity, film's thickness, and calculated effective deposition time and growth rate.

The crystalline structure of yttria stabilized zirconia films was analyzed by GIXRD. In thermodynamic equilibrium, pure zirconia changes its phase from room temperature monoclinic to tetragonal and cubic phases at high temperature. With the addition of Y_2O_3, cubic zirconia can be sChart in a wide range of temperatures, from room temperature up to its melting point; this characteristic of yttria stabilized ZrO_2 makes it of technological importance [28]. Films deposited by aerosol assisted CVD [17] were polycrystalline with single fluorite structure. Any phase segregation into Y_2O_3 or other compound was observed. Figure 5 shows the GIXRD pattern of different YSZ samples as a function of Y at. % in solution, for the 2θ interval between 27° and 37°; intensities were normalized to show the differences in peak position; the inset shows representative GIXRD pattern of YSZ film. The observed peaks are consistent with (1 1 1) and (2 0 0) planes of cubic structures; undoped sample (QA) to the ICDD card 00-065-0461 [29] and all doped films (QB to QE) to ICDD card 00-030-1468 [30], independently of Y content due to its relatively low values.

Table 4. Experimental aerosol velocity, film's thickness and calculated effective deposition time and growth rate

Sample	Droplet velocity [cm.s^{-1}]	Film Thickness [nm]	Effective deposition time [min]	Effective growth rate [nm.min^{-1}]
Z01	72	435	3.9	111
Z02	60	350	3.9	89
Z03	48	143	2.0	73
Z04	43	93	2.2	43
Z05	37	160	2.8	58
Z06	72	363	3.9	92
Z07	50	181	6.2	29
Z08	42	530	6.2	85
Z09	216	202	0.7	289
Z10	216	208	0.7	297
Z11	189	194	0.8	243

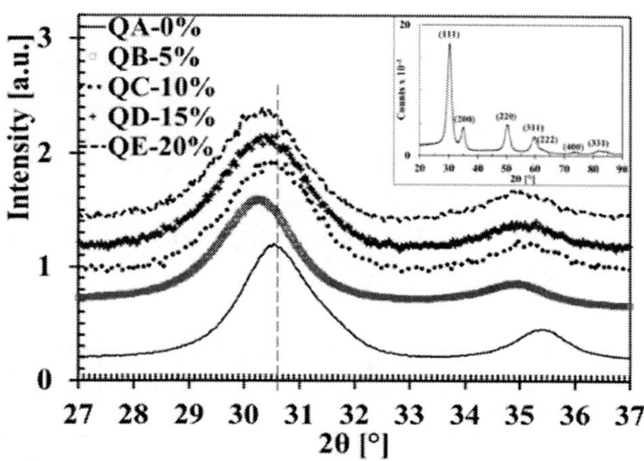

Figure 5. GIXRD pattern of different YSZ samples as a function of Y at. % in solution. Inset shows representative GIXRD pattern of YSZ film.

The peak position and broadening were analyzed by adjusting Gaussian curves. A change is shown in the (111) peak position for undoped (30.603°) and Y doped zirconia films (30.516°–30.282°); i.e., a systematic shift of peak maxima towards low angles is observed as the dopant content increases. These results correspond to a regular increase of interplanar distances; they are tabulated in Table 5.

Lattice parameters were calculated from the position of principal diffraction peaks. Value of lattice parameter "a" (see Table 5), was lower than corresponding bulk values of undoped [29] (0.5088 nm) and doped films [30] (0.5139 nm), evidencing the existence of structural strain in the films. For doped samples, a consistent increase of lattice parameter with Y content was observed from 0.5074 to 0.5113 nm, coincident with lattice parameter tendency reported in the International Center for Diffraction Data cards.

In addition, the analysis of the shape of x-ray lines can be used to determine crystallite or domain size using Scherrer's equation [31].

In the case of multilayered films, another contribution to the peak broadening can be the satellite peaks around the Bragg reflection, due to the periodic multilayered structure of the film [32]. The position of the mth order satellite peak ($\theta_{\pm}$) can be obtained from:

$$\sin(\theta_{\pm}) = \sin(\theta_B) \pm \frac{m \cdot \lambda}{2 \cdot p} \qquad (3)$$

where θ_B is the Bragg angle, λ is the wavelength of x-ray, and p is the period of the multilayer structure. However, any indication of satellite peaks was detected in the experimental GIXRD patterns and we then considered that peak broadening can only be caused by crystallite size. The experimental FWHM of the diffraction peaks was corrected with the instrumental broadening and assuming no strain broadening. Crystallite size was around 7 nm, consistent with TEM results, with a slight tendency to increase with Y content. They are tabulated in Table 5.

Table 5. Interplanar distances, lattice parameter and crystallite size of different YSZ samples as a function of Y at. % in solution

Film	d_{exp} [nm]	a [nm]	Crystallite size [nm]
QA-0%	0.2921±0.0001	0.5060±0.0001	6.7
QB-5%	0.2929±0.0001	0.5074±0.0001	6.7
QC-10%	0.2935±0.0001	0.5084±0.0001	6.6
QD-15%	0.2946±0.0001	0.5112±0.0001	7.0
QE-20%	0.2951±0.0001	0.5113±0.0001	7.2

The second group of analysed samples corresponds to ZnO based thin films, samples C, D, E, F and G in Table 2. Undoped and doped ZnO:d (d: Co, Sc, Mg) films were studied as a function of dopant content. After optimization of the deposition parameters, such as substrate temperature, carrier gas flux, nozzle velocity, etc. high quality, uniform, highly transparent, non-light scattering, and well adhered, polycrystalline thin films were obtained. In general dopant to Zn ratio in film did not coincide with that in solution [33]. Dopant content finally depends on the deposition rate of the dopant element compared with that of the host oxide.

These rates will be determined by several physico-chemical properties of the precursor compounds (dopant and host) and the deposition parameters. Table 6 resumes the comparison of experimental values of d/Zn at. ratio in film, measured by EDS, to that in solution for several dopant elements in ZnO. It is interesting to state that the same type of precursor (acetate) gave a very different ratio dopant/Zn in the film compared to that in solution; it varied from 0.33 (Mg, Cu) to 1.5 (Co).

It is worthwhile to remark that in general solubility limit in ZnO, of many elements such as In [34], Al, V, Sc, etc. is low, i.e., only a few at. %. However Co and Mg are exceptions; since very high solid solubility of CoO in Wurtzite ZnO, can be metastably extended up to 40 at.% [35], and Mg solubility, as high as 33% [36], was reported. If dopant content surpasses its solubility limit, it starts to segregate and develop a different compound or crystalline phase.

Secondary electron SEM images of undoped and doped ZnO films on BSG substrates are shown in Figure 6. These SEM micrographs show that the surface morphology of the films is strongly dependent on the type of the dopant atoms, as reported elsewhere [33]. Figure 6a shows typical morphology of undoped ZnO film, composed of flake-like grains, parallel to basal plane of hexagonal ZnO structure; the flakes are inclined in relation to the substrate surface. Figure 6b presents the surface of Co doped sample; it is shown that the film's microstructure was constituted of close packed globular grains of different sizes composed of many small crystallites. Surface morphology of Sc doped film is shown in Figure 6c; numerous compact faceted triangular grains can be seen as constituents of the film. For Mg doped sample (Figure 6d), its superficial microstructure was a combination of many flake-like grains intercalated with almost globular grains of different sizes.

The cross section of a bi-layer coating of ZnO and Co doped ZnO (sample D) deposited onto borosilicate glass substrate (BSG). A ZnO buffer layer was deposited to avoid diffusion of glass elements, principally Na and Si, into ZnO:Co active layer, used for photocatalytic purposes. The composition and elemental distribution of the different layers can be obtained by EDS microanalysis and related modes, such as line scanning and mapping.

Table 6. Ratio of experimental values of dopant/Zn at. in film to that in solution. Dopant content in film was measured by EDS. ZnO precursor was Zn acetate. The dopant precursor is also tabulated

Dopant	Precursor	d/Zn at. film/solution
Co	Co (II) acetate	1.5 ± 0.1
Sc	Sc (III) acetylacetonate	0.34 ± 0.04
Mg	Mg (II) acetate	0.33 ± 0.06
In	In (III) acetate	1.0 ± 0.1 [33]
Cu	Cu (II) acetate	0.33 ± 0.06 [33]

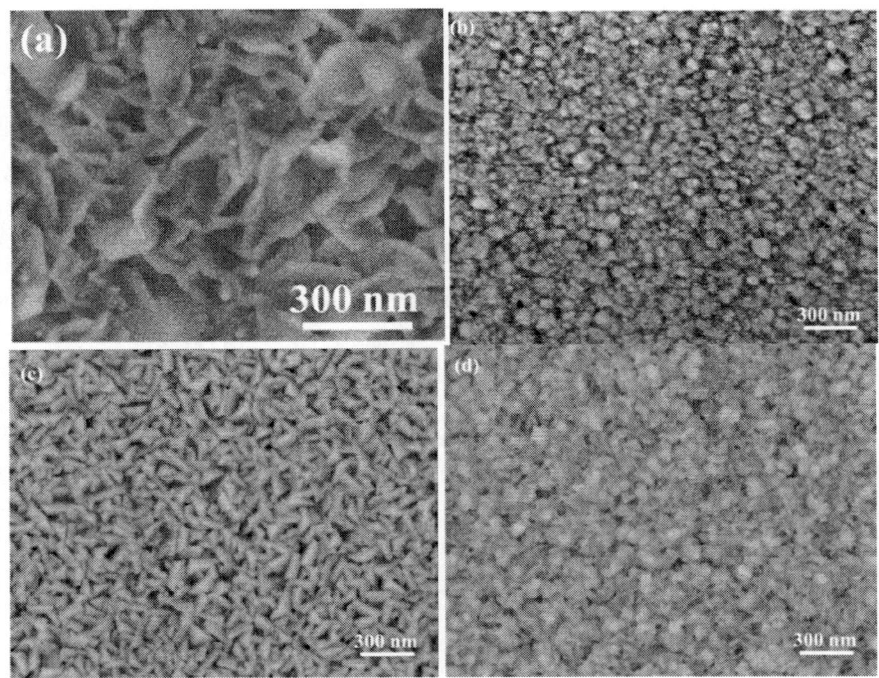

Figure 6. Secondary electron SEM micrographs of undoped and doped ZnO thin films. a) undoped; b) 10% Co doped; c) 20% Sc doped and d) 5% Mg doped.

Figure 7a shows a bright field TEM overview of the cross section; a total thickness for the bi-layer of around 140 ± 6 nm was obtained. Elemental analysis by EDS indicates that no contamination was present in the film.

In addition, film stoichiometry was close to ideal ZnO, and doped films maintained approximately the same Co/Zn ratio in film as in solution [18].

Figure 7b presents EDS line scanning result showing the distribution of characteristic x-ray counts of each element present in the coating. Elemental distribution is qualitatively related with these counts.

Thickness of each layer can be deduced from this distribution, i.e., a ZnO buffer layer of around 34 ± 3 nm and ZnO:Co layer of 104 ± 3 nm. A sharp interface of less than 5 nn was observed between both layers. A decrease of Zn counts is observed towards the surface of the film.

Similar results are found in the pertaining documents; see for example P. Amézaga-Madrid et al. for the case of YSZ multi-layered thin film [27].

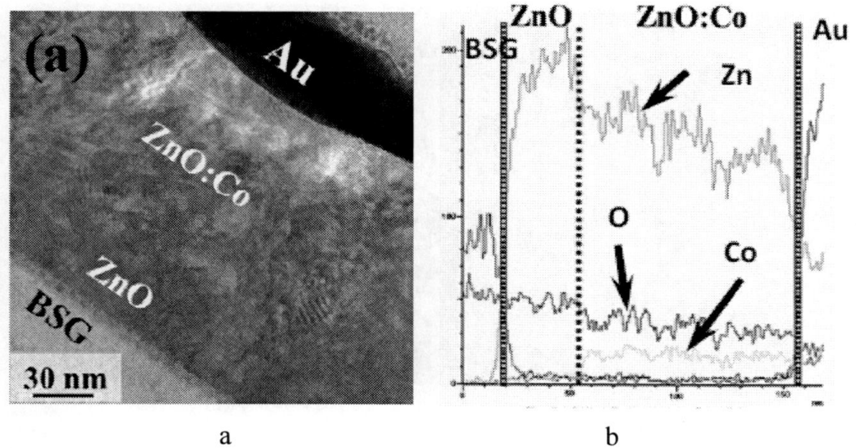

Figure 7. Elemental analysis of the cross section of a bi-layer coating (sample D). a) Bright field TEM images of the cross section. One can also see the BSG (bottom layer), ZnO and ZnO:Co, and Au layer deposited for protection during FIB preparation. b) Line scan results showing the distribution of the elements in the bi-layer. Other distributions not identified correspond to Si in BSG and Au deposit.

In HRTEM observations, films are mainly constituted by many close packed nanoparticles. On closer inspection, recurrent values of interplanar distance between lattice planes were found, for example 0.253 ± 0.003 nm (Figure 8a), which corresponds to {101} planes of the hexagonal structure of the ZnO wurtzite phase. All the data above supported the idea that undoped and Co doped ZnO were present in a single phase belongs to the hexagonal wurtzite structure. Additionally, high resolution TEM micrographs gave lattice fringes images(atomic planes) of the crystalline structure; analysis of these micro-graphs allows the measurement of inter-planar distances and determination of the family planes existent in the grains of the film. Figure 8 presents bright field HRTEM images of sample B. Figure 8a shows a sharp boundary between borosilicate glass substrate and ZnO layer; crystalline arrangement of planes (101) of ZnO grew immediately almost parallel to the glass surface. GIXRD analysis of undoped and doped ZnO shows that the films are polycrystalline, single phase with a Wurtzite type structure [37]. No other phases, corresponding to metallic dopant, other oxides or compounds were detected. A marked (002) texture for films of less than 200 nm of thickness is shown; this texture diminishes as the films thickness increases.

Figure 9 presents typical GIXRD patterns of films as a function of dopant content in solution.

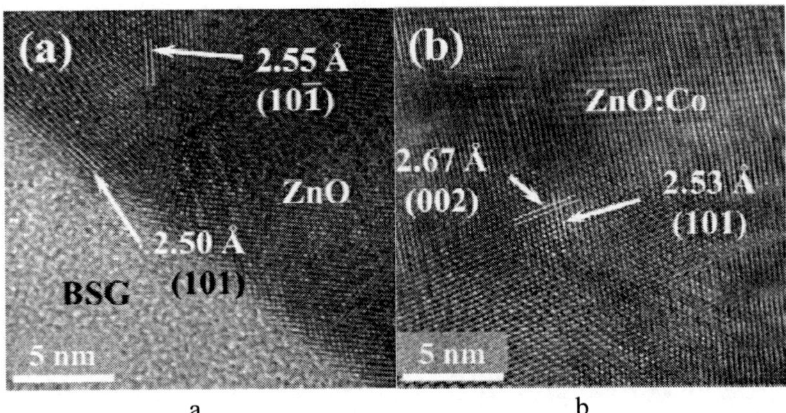

a   b

Figure 8. Bright field HRTEM images of bi-layer ZnO – ZnO:Co. a) sharp boundary between substrate and ZnO layer. b) lattice fringes of (101) and (002) family planes in ZnO:Co layer.

In the case of low Co concentration (Figure 9a), a slight decrease of diffraction lines position as Co content increases is shown. Consequently, interplanar distance and lattice parameters increase, probably due to the existence of Co^{+3} ions that distort the lattice structure, creating defects. On the other hand, for high Co concentration all patterns are very similar, without any peak shift, as depicted in Figure 9b. Figure 9c presents GIXRD patterns of Sc doped films. A similar shift of diffraction lines position to low angles as dopant content increases is shown, with the consequent interplanar distance and lattice parameters increase. For Mg doped ZnO films, a more significant shift was observed, but in the opposite direction, i.e., the diffracting line position increases with the increase of Mg. Table 7 presents the results of interplanar distance and lattice parameter "c" calculated from diffraction lines positions. For Co doped films, lattice parameter "c" initially increases with Co content, up to around 3 at. %, then it decreases and stay almost constant as Co concentration augment up to 20 at. %. This behavior is consistent if at low concentration, only Co^{+2} ions are incorporated in the Zn sites, then the in-plane atomic array will be rather reduced. Consequently, lattice axis parallel to in-plane should become smaller and the c-axis becomes larger. For higher concentration, both Co^{+2} or Co^{+3} remain at interstitial sites as well as in Zn site of ZnO. Consistent with Urbach tail observed in absorption spectra, as discussed in the optical properties section of this chapter. Disturbing the in-plane assembly of atoms, increasing the a-axis lattice parameter and decreasing the c-axis parameter [38].

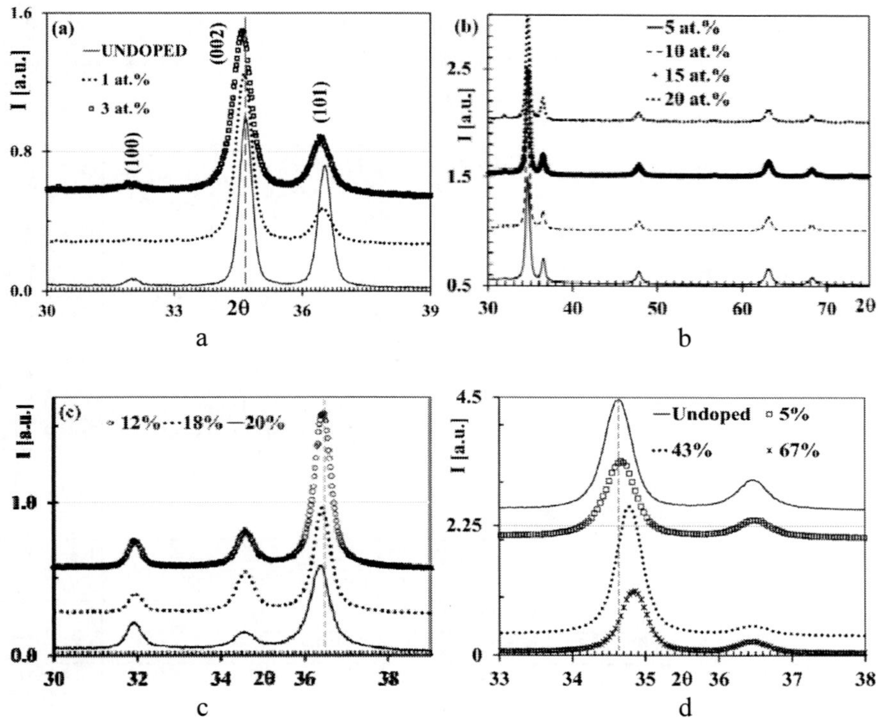

Figure 9. GIXRD patterns of undoped and doped ZnO films as a function of dopant content in solution. a) low Co concentration; b) high Co concentration; c) Sc dopant; d) Mg dopant. Dashed line indicates the position of diffraction line of undoped sample.

Analysis of peak broadening in GIXRD patterns allowed us to calculate crystallite size, assuming no strain broadening, employing Scherrer's equation and correcting the FWHM of the diffraction lines with the instrumental broadening. Results are presented in Figure 10.

Instrumental broadening around the (100), (002) and (101) peaks of ZnO was measured using a Si standard, we obtained a 2θ broadening of $0.2°$.

For Co dopant, a decrease of crystallite size as the Co content increases up to 0.05 Co/Zn at. ratio was found, then a slight increase was observed as the Co concentration continued increasing (see Figure 10a).

For Sc doped films, GIXRD patterns [19] present similar characteristics as those of Co doped ones; however crystallite size decreases with the increase of Sc concentration, up to around 0.08 Sc/Zn at. ratio (see Figure 10b).

Table 7. Interplanar distances, lattice parameter c, crystallite size and compressive stress σ of undoped and doped ZnO samples as a function of dopant/Zn at. ratio in solution

Sample	Co/Zn at. ratio solution	Co/Zn at. ratio film	Interplanar distance (002) [nm]	Lattice parameter c [nm]	Crystallite size [nm]	σ [Gpa]
C0	Undoped	-	0.2588±0.0001	0.5175±0.0001	48±1	2.8±0.1
C01	1%	ND	0.2591±0.0001	0.5182±0.0001	37±1	2.1±0.1
C03	3%	ND	0.2593±0.0001	0.5185±0.0001	29±1	1.9±0.1
C05	5%	8%	0.2588±0.0001	0.5175±0.0001	26±1	2.8±0.1
C10	10%	16%	0.2589±0.0001	0.5178±0.0001	29±1	2.5±0.1
C15	15%	26%	0.2588±0.0001	0.5175±0.0001	30±1	2.8±0.1
C20	20%	43%	0.2588±0.0001	0.5175±0.0001	34±1	2.8±0.1
Sample	Sc/Zn at. ratio solution	Sc/Zn at. ratio film	Interplanar distance (002) [nm]	Lattice parameter c [nm]	Crystallite size [nm]	σ [Gpa]
S65	20%	6.5%	0.2595±0.0003	0.5192±0.0005	24±1	1.3±0.1
S73	20%	7.3%	0.2598±0.0003	0.5197±0.0005	26±1	0.9±0.1
S76	20%	7.6%	0.2597±0.0003	0.5195±0.0005	27±1	1.0±0.1
S53	18%	5.3%	0.2594±0.0003	0.5189±0.0005	36±1	1.6±0.1
S43	12%	4.3%	0.2594±0.0003	0.5188±0.0005	33±1	1.6±0.1
Sample	Mg/Zn at. ratio solution	Mg/Zn at. ratio film	Interplanar distance (002) [nm]	Lattice parameter c [nm]	Crystallite size [nm]	σ [Gpa]
M0	Undoped	-	0.2591±0.0001	0.5182±0.0001	35±1	2.2±0.1
M05	5%	2%	0.2589±0.0001	0.5177±0.0001	30±1	2.6±0.1
M43	43%	13%	0.2580±0.0001	0.5161±0.0001	38±1	4.0±0.1
M67	67%	22%	0.2575±0.0001	0.5149±0.0001	34±1	5.0±0.1

The decrease of crystallite size can be explained by the creation of more nucleation centers as the dopant quantity increases at low concentration, causing the reduction of crystallite size.

For Mg doped films a tendency to decrease the lattice parameter "c" with Mg content was observed (see Table 7), coincident with trend reported by M. Yano et al. [39] for $Mg_xZn_{1-x}O$ films deposited by MBE.

A linear fit was calculated for the lattice parameter "c" as a function of Mg content x in the wurtzite structure of the film; the obtained slope was very close to that reported by Yano (-0.017) [39]:

$$c = 0.5204 - 0.0147x \tag{4}$$

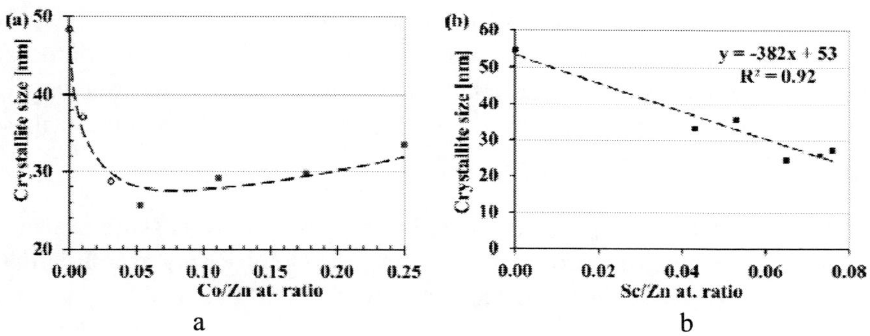

Figure 10. Crystallite size of undoped and doped ZnO thin films as a function of dopant concentration. a) Co doped; b) Sc doped.

In addition, changes in lattice parameter c were utilized to estimate the stress in the films following the relation:

$$\sigma\,[Pa] = -453.6 \cdot 10^9 \left(\frac{c - c_0}{c_0} \right) \tag{5}$$

A compressive stress in the direction of c axis was obtained for all the films (see Table 7): Co (1.9 – 2.8 GPa), Sc (0.9 – 1.6 GPa) and Mg (2.2 – 5.0 GPa).

3.3. Microstructural Characteristics of Nanostructured Oxides Obtained by Aerosol Assisted CVD

One example of nanostructured oxides, synthesized by AACVD, is the spherical hollow magnetite nanoparticles reported by B. Monárrez et al. (sample K). Figure 11a shows typical magnetite nanostructures with spherical hollow geometry, composed of hundreds of adjacent crystallites forming the crust. Morphological analysis of sample deposited at 723 K, shows that average external particle diameter was around 300 ± 78 nm, with a crust thickness of 26 ± 5 nm, crystallite size of 19 ± 5 nm and porosity of 30 % (see Figure 11b). Slight influence of synthesis temperature in the morphological characteristics of the magnetite nanoparticles was found (see Table 8).

Average aerosol droplet diameter was estimated by optical interferometry to be around 2.2 ± 0.2 μm.

Considering the precursor contained in each droplet of average size, the average diameter of a hollow nanostructure was calculated. It was confirmed that one aerosol droplet generated only one hollow nanostructure of approximately the same diameter, crust thickness and porosity of those obtained experimentally.

Table 8. Average diameters of particle, crust thickness, crystallite size, porosity, experimental and theoretical specific surface area as a function of synthesis temperature in an AACVD process

Temperature [K]	Average particle diameter [nm]	Average crust thickness [nm]	Average crystallites size [nm]	Porosity [%]	Specific surface area [m^2.g^{-1}]	Theoretical specific surface area [m^2.g^{-1}]
723	300±78	26±9	19±5	30±5	66±3	63
773	380±100	27±5	25±6	59±5	47±3	48

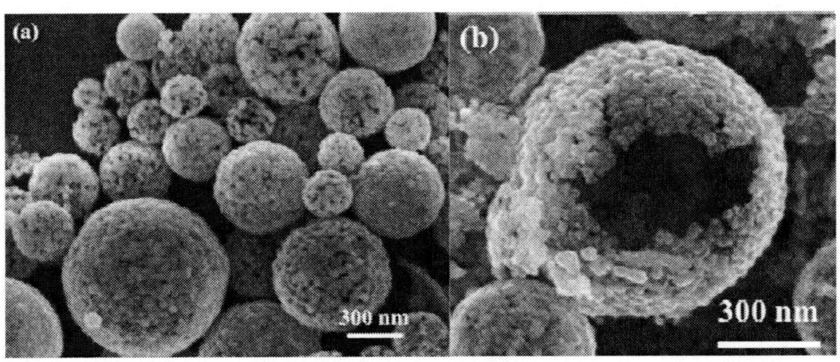

Figure 11. Secondary electron SEM micrograph of typical hollow spherical magnetite nanostructures obtained by AACVD. a) Overview of porous spherical nanostructures. b) Broken spherical nanoparticle showing the hollow structure.

The experimental specific surface area was around 62 ± 4 m^2.g^{-1}, in accordance with the one calculated considering an isolated spherical nano-particle of diameter equal to the average crystallite size. This means, that the porous microstructure of the hollow nanoparticles, causes the component crystallites to perform as isolated spherical nanoparticles in regard to surface related phenomenon.

Then, hollow spherical magnetite nanoparticles (~300 nm of diameter) of this work have advantageous characteristics concerning particle manipulation and recollection and, on the other hand, the component (agglomerated) crystallites (~20 nm of size) perform as isolated spherical nanoparticles with regard to their specific surface area.

The formation of these nanostructures has been studied by several authors [40, 41]; it was explained by the following mechanism: considering the transport of monodisperse aerosol droplets of fixed diameter, solution concentration and flow rate into a tubular furnace reactor, maintained at a uniform and constant temperature.

As the aerosol droplet comes into the furnace, it heats and starts to evaporate. As the droplet heating continues, its size decreases and the solution concentration in the droplet increases, resulting in a higher concentration on the surface rather than in the interior. At some stage, surface concentration may reach critical supersaturation and solute nucleation takes place.

Depending on the evaporation rate of the solvent and diffusion rate of the solute, the concentration at the interior of the droplet can be higher or lower than the equilibrium saturation at the time of nucleation. If the diffusion rate of the solute (compared) to the interior of the droplet is adequately high or evaporation rate sufficiently low, its concentration in the interior can be higher than equilibrium saturation. In this case, solute nucleation occurs in the entire droplet and a solid particle forms. In the opposite case, if solute concentration in the interior is lower than equilibrium saturation, solute nucleation takes place only on the surface, forming a crust and a hollow particle.

Furthermore, the remnant solvent has to evaporate through the crust if it is sufficiently porous; if not, the pressure of solvent vapour can break the particle.

Many experimental conditions, solvent and solute properties affect the evaporation rate of the solvent and the diffusion rate of the solute; consequently, they affect the distribution of solute concentration inside the droplet and determine the formation of solid or hollow nanostructures.

While, evaporation rate depends mainly on furnace temperature, carrier gas flow, solvent boiling temperature, solvent heat of evaporation, solvent vapour pressure and solution concentration; diffusion rate mainly depends on droplet temperature, solute diffusion coefficient, and solution concentration.

Table 9 resumes the principal parameters and how their magnitude determines the morphology of the nanostructure.

Figure 12 shows GIXRD patterns of magnetite nanoparticles obtained at 723 and 773 K. It shows the characteristics diffracting lines of magnetite, identified by JCPDS card 01-089-0691 [42].

In addition, magnetite phase was confirmed by Raman spectroscopy, discarding a possible maghemite phase, which has similar diffraction pattern than magnetite. Crystallite size estimated by Scherrer equation gave coincident values (~20 nm) to that measured by electron microscopy.

Table 9. Principal experimental conditions and properties that determine the formation of solid or hollow nanostructures

Parameter	Solid	Hollow
Experimental conditions		
Synthesis temperature	Low	High
Solution concentration	High	Low
Carrier gas flow	Low	High
Solvent properties		
Boiling temperature	High	Low
Heat of evaporation	High	Low
Vapour pressure	Low	High
Solute properties		
Diffusion coefficient	High	Low
Saturation concentration	Low	High

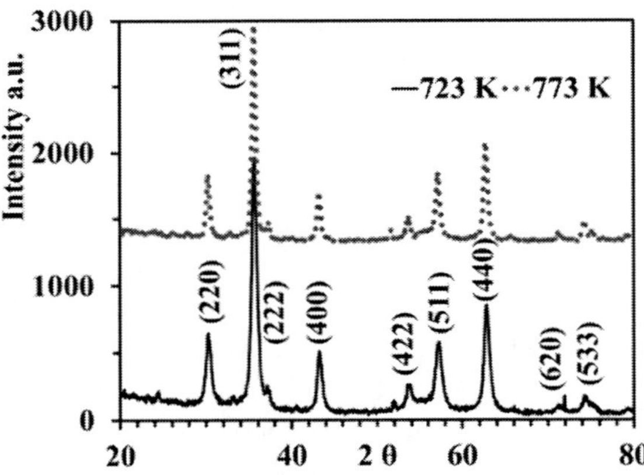

Figure 12. GIXRD patterns of magnetite nanoparticles obtained at 723 and 773 K.

4. MICROSTRUCTURE PROPERTY RELATIONS

4.1. Optical Properties of Thin Films Deposited by Aerosol Assisted CVD. Correlation with Microstructural Characteristics

Optical properties were evaluated by total transmittance and reflectance, direct transmittance and near normal absolute reflectance spectra using a Cary 5000 UV-visible-NIR spectrophotometer in the range of 300–2500 nm.

Total transmittance and reflectance were acquired using a diffuse reflectance DRA 2500 accessory. The films deposited onto fused silica substrates were used for these analyses. Optical constants (n and k) of the films were obtained from direct transmittance and near normal absolute reflectance spectra.

TFcalc v.3.5.14 software was employed to fit the simulated spectra of film-substrate stack to the experimental data. Considering phase incoherency in the substrate [43], the refractive index n_s of the fused silica substrate, was modeled with a Sellmeier [44] dispersion relation.

On the other hand, extinction coefficient k_s of fused silica was taken as zero, in the wavelength range utilized (200–1000 nm). For the optical constants of the films, we have employed Sellmeier and exponential dispersions, for $n(\lambda)$ and $k(\lambda)$, respectively [43]. The Sellmeier and exponential parameters were obtained using TFcalc software, by fitting the simulated and experimental spectra.

$$n(\lambda) = \sqrt{A_0 + \frac{A_1 \lambda^2}{\lambda^2 - A_2}} \text{ and } k(\lambda) = B_1 \cdot exp\left(\frac{B_2}{\lambda}\right) \tag{6}$$

Absorbance $A(\lambda)$ spectra were calculated from total transmittance $T_t(\lambda)$ and reflectance $R_t(\lambda)$ measurements using:

$$A(\lambda) = 1 - T_t(\lambda) - R_t(\lambda) \tag{7}$$

Absorbance values were correlated with photocatalytic activity of the films.

Absorption coefficient α spectra were obtained from optical constant k, deduced from exponential dispersion parameters obtained by TFcalc modeling of experimental direct transmittance and absolute reflectance spectra. Values of α higher than 10^5 cm^{-1} were obtained in the absorption region, which are typical for interband transitions.

Considering the efforts to design and develop optoelectronic devices, band gap engineering is a crucial step, being this is of particular importance for ZnO based materials. It consists of a combination of a host semiconductor with another of different band gap, resulting in a material where its band gap can be adjusted accurately.

Therefore, an exact method for the determination of the energy band gap is necessary; here an optical method is implemented.

Band gap energy of several films was estimated using the well-known Tauc's relation [45], between absorption coefficient and photon energy. Thus, the following relation exists in an energy interval, close to the band gap energy:

$$\alpha(\lambda) \cdot E = B \cdot (E - E_g)^p \tag{8}$$

where $\alpha(\lambda)$ is the absorption coefficient, E is the photon energy, B is an energy independent quantity, E_g is the optical band gap energy and p is the exponent that depends of the type of transition. For a permitted direct transition p=1/2. Then, the linear portion of the $[\alpha(\lambda) \cdot E]^{1/p}$ plot as a function of the energy intercepts the energy axis at the optical band gap energy E_g.

Packing density δ_p, of samples can be estimated using effective refractive index n in an equivalent expression to that reported by Sharma [46].

$$\delta_p = \frac{n^2-1}{n^2+1} \cdot \frac{n_B^2+1}{n_B^2-1} \tag{9}$$

where n is the refractive index of the film and n_B is the bulk refractive index of the material. Several equations are proposed and some discrepancy exists around which is the most accurate to estimate the porosity in thin films from optical transmittance data.

Other authors propose that a more accurate expression is given by [47]:

$$\delta_p = \frac{n^2-1}{n_B^2-1} \tag{10}$$

Several zirconia based thin films were optically characterized for determination of optical constants and band gap energy. High optical quality of the films was observed since they were non-light scattering (diffuse reflectance and transmittance low) and had high visible transmittance. Optical properties were correlated with microstructural characteristics of the samples, such as dopant content, porosity, structural defects, etc.

Table 10 presents the Sellmeier and exponential parameters obtained for representative zirconia based films and fused silica substrate. These parameters were used to analyze the optical properties of zirconia-based thin films; the results are presented below. Figure 13a displays the transmittance and reflectance spectra of representative undoped ZrO_2 thin films compared to those simulated by TFcalc.

Table 10. Parameters of Sellmeier and exponential dispersions of optical constants of undoped and yttrium doped zirconium oxide thin films deposited onto fused silica substrates

Sample	Sellmeier			Exponential	
	A_0	A_1	A_2	B_1	B_2
Fused silica	2.30	$4.1777.10^{-1}$	$3.90.10^{-2}$	0	0
QA-0%	2.8237	$9.5834.10^{-1}$	$3.3780.10^{-2}$	$9.4876.10^{-7}$	2.5382
QB-5%	2.8076	$8.3014.10^{-1}$	$3.3727.10^{-2}$	$9.0841.10^{-7}$	2.5470
QC-10%	2.6798	$8.2785.10^{-1}$	$3.4193.10^{-2}$	$3.1429.10^{-8}$	3.2645
QE-20%	2.6484	$8.2182.10^{-1}$	$3.4007.10^{-2}$	$2.6575.10^{-7}$	2.7930

Very good fitting between experimental and simulated spectra is shown in Figure 13a, using the Sellmeier and exponential parameters of Table 10.

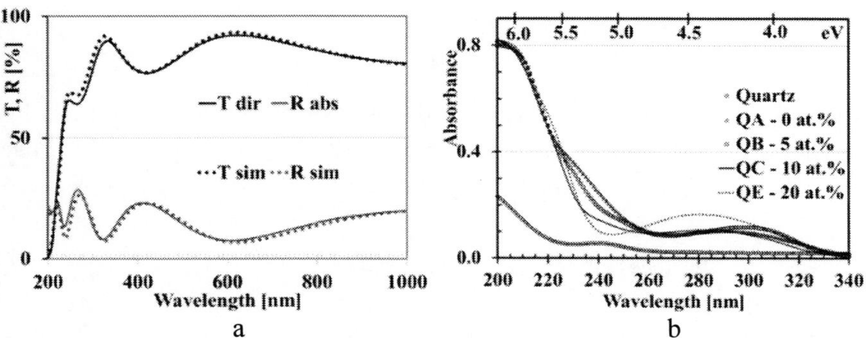

Figure 13. a) Experimental direct transmittance Tdir and absolute reflectance Rabs spectra of undoped ZrO_2 thin films compared to those simulated by TFcalc. b) Absorbance spectra of fused silica substrate, undoped and Y doped ZrO_2 thin films.

Absorbance spectra in the UV region were obtained from total transmittance and reflectance measurements; they are presented in Figure 13b for undoped, Y doped films and fused silica substrate.

A fine structure in the absorption spectrum can be observed similar to that reported for ZnO [48]; it can be attributed to localized states below the conduction band gap edge, as discussed below.

In addition, results of refractive index n calculated by Sellmeier dispersion relations are depicted in Figure 14a. It is shown that n is almost constant in the near IR, it increases less than 3% in the visible region and finally increases significantly more pronouncedly in the UV.

In general, a decrease of n is found as the Y content in the film increases; this behavior can be justified considering that oxygen vacancies are incorporated in the zirconia lattice proportionally to Y concentration, which lowers the material density and consequently the refractive index [49]. In the visible interval, an almost constant value of n of 2.01 was obtained for undoped ZrO_2 films; it decreased to around 1.93 for Y doped with 20 at. %.

Accordingly, packing density δ_p was calculated using equation (9) and considering 2.17 as the bulk refractive index of YSZ [47]; values around 0.93 – 0.90 were obtained for analyzed zirconia based films (see Table 11). Equation (10) gives lower values of packing density, around 0.80.

Another semi-empirical dispersion relation of refractive index, which is related with parameters associated with the structural order of the material, is the Wemple-Di Dominico equation [50]. In this approach, the dispersion of the refractive index was described with the single oscillator model according to:

$$n^2 = 1 + \frac{E_d E_0}{E_0{}^2 + (h \cdot v)^2} \tag{11}$$

where E_d, the dispersion energy, is a measure of the average strength of interband optical transitions; E_0 is the oscillator energy; h is the Planck constant and v is the frequency. Dispersion energy can be evaluated by:

$$E_d = \beta \cdot N_c Z_a N_e [\text{eV}] \tag{12}$$

where β is a parameter that depends on the interatomic bond, N_c is the effective coordination number of the cation nearest-neighbor to the anion, Z_a is the chemical valence of the anion, N_e is the effective number of valence electrons per anion excluding the cores.

Table 11. Calculated values of packing density δ_p, optical band gap E_g, intensity of (111) diffracting line, dispersion energy E_d, oscillator energy E_0, and parameter β of dispersion energy

Sample	δ_p	E_g [eV]	$I_{(111)}$	E_d	E_0	β
QA-0%	0.93	5.93	1508	23.9	9.13	0.37
QB-5%	0.92	5.94	1681	25.5	9.86	0.40
QC-10%	0.90	5.99	1640	23.5	9.59	0.37
QE-20%	0.90	6.00	1575	23.2	9.60	0.36

The dispersion energy E_d is associated with the changes in the structural order of the material; the more ordered the materials the larger E_d [51]. E_d and E_0 were evaluated by plotting $\frac{1}{n^2-1}$ as a function of photon energies $(h.v)^2$, below the interband absorption edge. Linear least square fitting of this plot relates the slope to $\frac{1}{E_d E_0}$ and the intercept to $\frac{E_0}{E_d}$. Results are tabulated in Table 11.

One can observe that E_d and E_0 increased for 5 at. % of Y sample, then they decreased at higher Y content. The highest value of E_d for YSZ film with 5 at. % of Y correlates with the highest intensity of (111) diffracting line (see Table 11) and hardness and elastic modulus, reported elsewhere [17].

Moreover, parameter β was estimated from the calculated values of E_d considering for zirconia films: $N_c = 8$ [52], $Z_a = 2$, and $Ne = 4$. Values of β around 0.37 were obtained, which indicate the covalent character [53] of interatomic bonds in agreement with calculations reported by Nakamatsu et al. [54].

However, some controversy exists around the bond character in zirconia, because it is considered also an ionic crystal. Finally, values of β will depend on the effective number of valence electrons per anion considered.

Absorption coefficient α was obtained from optical constant k; values of the order of $10^5 - 10^6$ cm^{-1} are typical for interband transitions. In the case of zirconia, it corresponds to transitions from the O 2p valence band to the Zr 4d conduction band.

Optical band gap energy was determined for zirconia based films in Figure 14b; it is found E_g in the range of 5.93 – 6.0 eV (see Table 11).

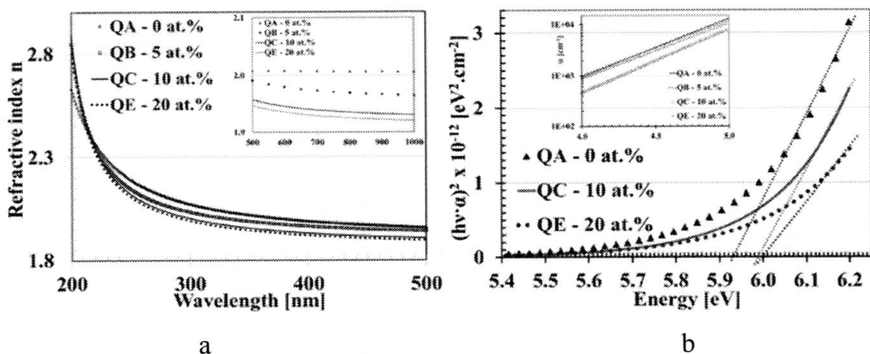

a   b

Figure 14. a) Refractive index of undoped and Y doped ZrO_2 thin films. Inset shows extended wavelength interval. b) Tauc's plot for the determination of optical band gap of undoped and Y doped ZrO_2 thin films; sample QB-5 at.% not shown for clarity. Inset shows semi-logarithmic plot of α in the energy range below the band gap (4 – 5 eV); it is shown a linear dependence, indicating an Urbach-type behavior.

These high values agree well with optical band gap data published previously for crystalline YSZ thin films with small crystallite size, of the order of 10 nm [55], consistent with the effect of quantum confinement. The different slopes suggest a different distribution and density of the electronic states.

The increase of the band gap energy is consistent with the observed decreases of refractive index as Y content increases in the films. There are various empirical or semi-empirical relations between n and E_g [53]; for example the Hervé-Vandamme correlation:

$$n^2 = 1 + \left(\frac{A}{E_g + B}\right)^2 \tag{13}$$

where A $\approx$ 13.6 eV and B $\approx$ 3.4 eV are constants.

Large discrepancy among reported optical band gap energy data covers a range of ~3.5 [47] to 6 eV. Low value reports probably originate from a misinterpretation of the increase in absorption at photon energies bellow the band gap (3 – 5 eV) as an interband transition.

Heiroth et al. [49] propose an energy level diagram for YSZ taking into account the fundamental interband transition (E_g) from the O 2p valence band to the Zr 4d conduction band and localized states in the gap 0.73 eV below the conduction band edge.

In Figure 13b, the fine structure observed in the absorbance spectra for energies below 5.5 eV, suggests the presence of those localized states. These states are generated by the extrinsic defects introduced by Y dopant, and are the responsible for the absorption at photon energies lower than E_g.

In addition, inset in Figure 14b shows a semi-logarithmic plot of absorption coefficient α as a function of photon energies, in the energy interval below the band gap; the linear dependence reveals an Urbach-type behavior [56]. This behavior is attributed to a structural disorder induced by the Y dopant.

The second group of optically studied films corresponds to zinc oxide based films. Optical properties are of singular interest for many applications of ZnO, here several methods are introduced to analyze them, with special attention to its relation with microstructural characteristics of the films.

Transmittance and reflectance of the different undoped and doped ZnO films were measured for determination of its absorption coefficient, band gap energy and dopant related absorption bands. Dopants inserted in the ZnO matrix were Co, Sc and Mg.

Figure 15a presents the direct transmittance and the absolute reflectance spectra of selected undoped and Co doped ZnO thin films. All the films have almost the same thickness, around 360 ± 30 nm. High optical quality of the films was observed since they were non-light scattering and had high visible transmittance.

Optical properties were correlated with microstructural characteristics of the samples, such as dopant content, porosity, structural defects, etc.

For Co doped samples, typical absorption bands in the visible region, due to Co ions, are present in the spectra. These bands give the characteristic slight cobalt blue color of the films, which becomes deeper as the Co concentration increases.

Reflectance spectra show modulation fringes indicating good quality films, i.e., low roughness interfaces, small absorption or dispersion and uniform thickness.

Absorption coefficient α was calculated from total transmittance and reflectance measurements. Figure 15b shows α as a function of photon energy E ($\log_{10}$ scale) and corresponding wavelength.

Different absorption processes in the Co doped ZnO films can be identified (see Figure 15b):

a) E_g fundamental interband absorption, consequence of electron excitation from the valence to the conduction band. This absorption is distinguished by the high absorption coefficient, larger than 10^5 cm^{-1}.

b) Co^{+2} absorption caused by dopant ions in the ZnO structure (discussed below), two bands around 1 and 2 eV [38] are recognized.

c) FCA free carrier absorption generated for the interaction of photons with free carriers in the material, with the subsequent loss of photon energy.

Sc and Mg doped ZnO films present similar transmittance and reflectance spectra.

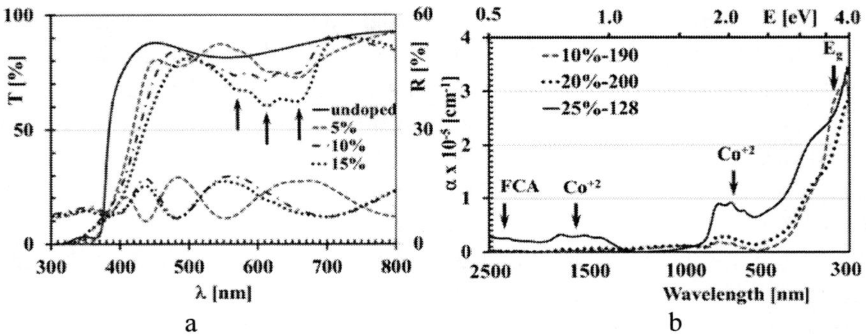

Figure 15. a) Total transmittance and reflectance spectra of undoped and Co doped ZnO thin films, for films of 360 ± 30 nm of thickness. b) Absorption coefficient α of representative Co doped ZnO films as a function of wavelength or photon energy E. The possible absorption processes are shown: E_g fundamental interband absorption, Co^{+2} due to dopant ions in the ZnO structure, and FCA free carrier absorption.

Fundamental interband absorption was studied for undoped and doped ZnO films for Co, Mg and Sc dopants, as a function of dopant concentration in solution and for two different film's thickness.

Optical band gap energy was calculated by Tauc's plot in Figure 16, results are presented in Table 12. It is shown a well-defined increase of E_g with dopant content for Co and Mg, higher also to that of undoped ZnO film.

Figure 17a shows Eg as function of dopant content x in the film, for Co and Mg doped samples. A linear correlation of E_g and dopant content in the film (x)is found, following the relations:

$$E_g^{Co}(x) = 3.30 + 0.5x \ and \ E_g^{Mg}(x) = 3.28 + 1.6x \ [eV] \tag{14}$$

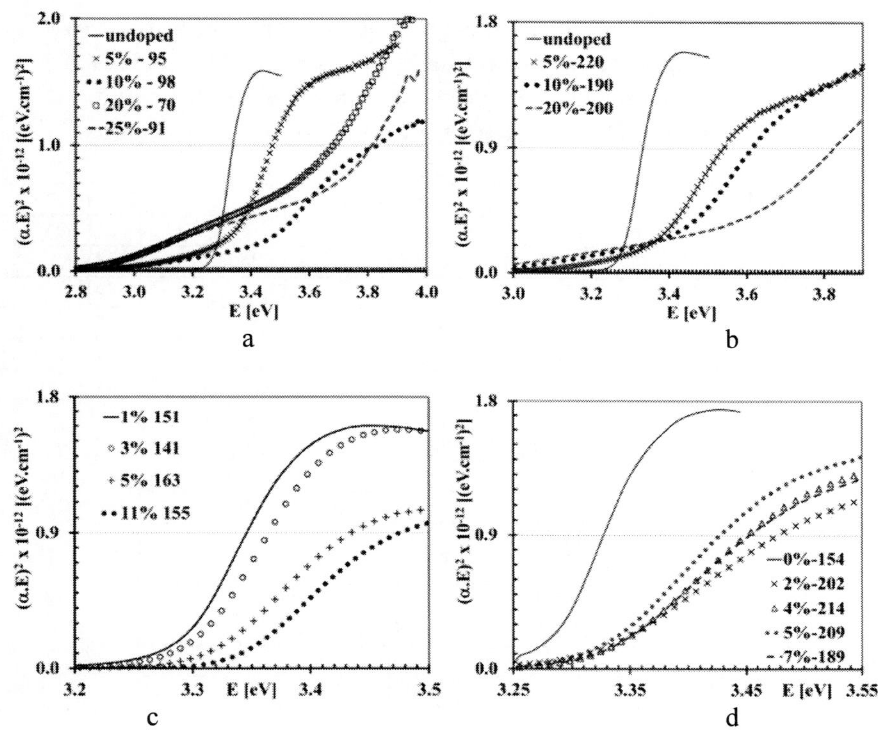

Figure 16. Optical band gap determination of ZnO based films by Tauc's plot. a) Co doped of 85 ± 15 nm of thickness. b) Co doped of 205 ± 15 nm. c) Mg doped of 149 ± 15 nm. d) Sc doped of 205 ± 12 nm. Co and Mg doped films show a significant increase of the band gap energy E_g as dopant content increases. In the case of Co doped samples a significantly increase of Urbach width ΔE was noticed as Co content increases; in contrast with the other dopants Mg and Sc.

In contrast, Sc doped films show an initial increase of E_g as dopant concentration attains 4 -5 at. %, then it stalls or tends to diminish (see Table 12).

Thinner samples seem to have slightly lower band gap energy compared to thicker Co doped films. Mg doped samples have almost no dependence of E_g on film thickness, only at high dopant content (> 40 at.%) a slight decrease of band gap energy is shown; probably caused by small variation in the dopant concentration in thinner and thicker samples. Sc doped films show small variations of E_g (± 0.005 eV), as their thickness changes between 130 to 190 nm.

Table 12. Optical band gap energy E_g of ZnO based thin films as a function of dopant concentration in solution and film (values in parenthesis) for Co, Mg and Sc dopants. Values of Eg and Urbach width ΔE for different film thicknesses are presented

Co/(Co+Zn) at.%		90 ± 11 nm		196 ± 14 nm	
solution	Film	E_g [eV]	ΔE [eV]	E_g [eV]	ΔE [eV]
0	0	3.28	0.08	3.28	0.08
5	8	3.32	0.40	3.33	0.32
10	16	3.39	0.76	3.39	0.65
20	43	3.46	1.16	3.52	1.21
25	-	3.55	2.06	ND	2.61

Mg/(Mg+Zn) at.%		86±12 nm	149±15 nm	
solution	Film	E_g [eV]	E_g [eV]	ΔE [eV]
0	-	3.28	3.28	0.08
1	-	3.29	3.29	0.10
3	-	3.30	3.30	0.06
5	2	3.31	3.31	0.05
10	-	3.34	3.33	0.05
20	-	3.47	3.44	0.14
30	13	3.50	3.50	0.19
40	22	3.64	3.60	-
50	-	3.67	3.72	-
60	-	3.77	3.80	-

Sc at.%	205 ± 12 nm	
solution	E_g [eV]	ΔE [eV]
0	3.28	0.08
2	3.31	0.30
4	3.33	0.32
5	3.33	0.29
7	3.32	0.30

For Co doped films, an important increase of Urbach width was observed. Urbach absorption tail behavior is generated by heavy doping of degenerate semiconductors, however it is also possible to be attributed to strong internal fields arising from ionized dopants or defects. Considering an exponential dependence of α on E, we have:

$$\alpha = \alpha_0 \cdot e^{\frac{E-E_0}{\Delta E}} \tag{15}$$

where α_0, E_0 and ΔE are parameters of a particular material. ΔE is the Urbach width, which is related to the extension of the band-tail into the band gap. ΔE was estimated by plotting $\ln(\alpha)$ vs E, for energies just below the band gap energy; the slope of the linear region corresponded to the reciprocal of ΔE. Figure 17b displays these plots for 5, 10, 20 and 25 at. % of Co, each sample was plotted in a particular interval below its band gap energy. Table 12 presents the values obtained as a function of dopant concentration. One can see that the Urbach width expands almost an order of magnitude, as Co content increases from 5 at. % to 25 at. %. This band tailing is attributed to point defects such as Co_{Zn}, interstitial Co^{+3}, and Zn vacancies [38]. Considering the absorption bands created by Co^{+2} ions (see discussion below) indicating Zn substitution by Co, the Urbach tail can be ascribed to Co_{Zn}.

In contrast Mg and Sc doped samples show small changes. Mg doped films present small values close or lower than undoped ones, with slight tendency to increase as the dopant content increases.

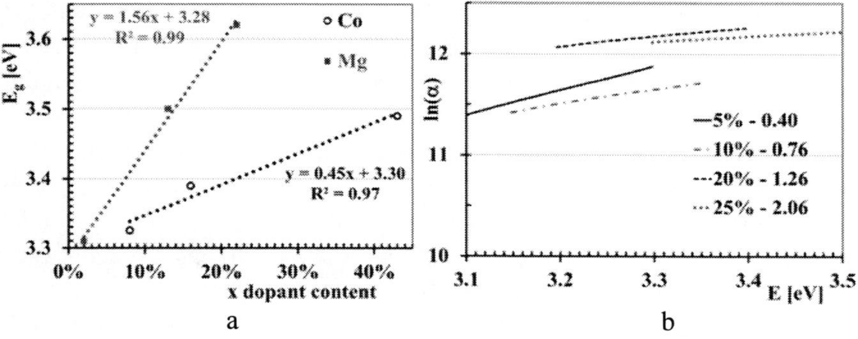

Figure 17. a) Optical band gap energy E_g as a function of dopant content x in film, for Co and Mg doped ZnO samples. b) Plot $\ln(\alpha)$ as a function of photon energy to determine the Urbach width ΔE of Co doped films, for several dopant concentrations in solutions. Obtained ΔE in eV is tabulated.

Sc doped samples have almost constant Urbach width of 0.30 ± 0.03 eV, for all dopant concentrations.

Ionic radius of 3d transition metals^{TM} depends on the valence state. However, when doped with 3d ions, they are in a valence state +2, ionic radius differences between Zn and 3d TM ions are minimized and defect formation for holding charge neutrality is suppressed.

Co ions could exist in either Co^{+2} or Co^{+3} states in ZnO. Then Co^{+3} states are expected to degenerate due to defect formation, owing to the higher valence state. Absorption edges associated with Co ions were found at longer wavelength, far from the fundamental interband absorption (see Figure 15b).

Co^{+2} ions generate characteristic absorption edges around 570 (2.19 eV), 620 (2.01 eV) and 660 (1.88 eV) nm [38, 57]. These bands are attributed to optical transitions between crystal-field-split levels of Co^{+2} ($3d^7$) ion. i.e., between intra-atomic *d-d* transitions of tetrahedrally coordinated Co^{+2} ions. For films of equal thickness, the intensities are expected to be proportional to the Co^{+2} concentrations.

Another IR band with three overlapped peaks at 0.75, 0.87 and 0.95 eV were observed, coincident with EELS measurement reported by A. Gorschlüter et al. [58]. The presence of these bands indicates that the Zn sites are occupied by Co^{+2} ions, leading to the suppression of defect formation by Co^{+3} ions.

Figures 18 show absorption coefficient α as a function of photon energy around the two Co^{+2} related bands. A comparison is shown between α for several ZnO films of different Co concentration; also shown is the influence of film thickness in sample with 25 at. % of Co. As expected magnitude of α increases as dopant content increases for films of similar thickness (200 nm); and also for fixed dopant concentration (25 at. %), α increases with the thickness of the sample.

These results are consistent with the calculated width of Urbach tail ΔE (see Table 12). Previously shown was an important increase of ΔE with Co concentration, coincident with an increase of the intensity of Co^{+2} related absorption bands, this correlation suggests that, in effect, Urbach tails are originated by occupation of Co^{+2} ions in Zn sites.

The last feature observed in the typical optical absorption spectra of Co doped ZnO films (Figure 15b) is the free carrier absorption (FCA) generated by photons of energy below E_g, which excite intraband transitions.

After the sharp drop in absorption with increasing wavelength, larger than that of the fundamental interband, a steady intraband absorption rise can be observed. Optical absorption can be dominated by FCA for photon wavelength larger than around 1 μm.

Classical theory predicts a λ^2 dependence for FCA; however experimentally a dependence is found between $\lambda^{1.5}$ and λ^4 [45], depending on the relevance of the particular scattering mechanism (acoustic phonon, optical phonon, impurity ion).

FCA results from the scattering of carriers in motion, due to disturbances of the crystal structure by phonons or impurities.

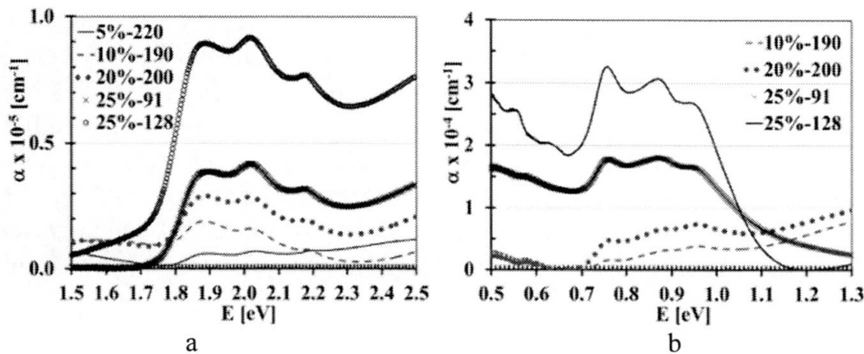

Figure 18. Absorption coefficient of Co doped ZnO of different dopant content. It is showing intra-atomic *d-d* transitions at 2 eV a); and 1 eV b).

In addition, FCA analysis in semiconductors has been constituted as an important characterization technique of its electronic properties with several advantages, over other electrical techniques, including being quantitative, contactless, and nondestructive.

Following a quantum mechanical treatment, in the high frequency limit ($v > 10^{13}$ s^{-1}), the free carrier absorption coefficient is commonly expressed with a wavelength dependence of the form:

$$\alpha_{FCA} = C \cdot \lambda^s \tag{16}$$

where the proportionality factor C depends on material parameters and the exponent s changes with the scattering mechanism.

The parameter C is proportional to the free carrier concentration N, which means that α_{FCA} increases with N; fact that is verified experimentally in several doped semiconductors. FCA was analyzed for Co and Sc doped ZnO films. Exponent s were estimated from the slope of the plot of $\ln(\alpha_{FCA})$ vs $\ln(\lambda)$, in a particular wavelength interval for each sample.

$$\ln \alpha_{FCA} = \ln C + s \cdot \ln \lambda \tag{17}$$

Figure 19 presents the calculated slope by linear least square fit of the plot of $\ln(\alpha_{FCA})$ vs $\ln(\lambda)$ in particular wavelength interval. For Co doped films two different wavelength intervals are shown over which FCA has typical power law dependence on wavelength. For thinner samples (*) around 90 ± 11 nm, the considered wavelength interval was 1.0 - 1.2 μm, and the obtained average s was of 4.2 ± 0.6; this value coincides with that reported by Kireev for impurity ions scattering [44]. For thicker samples (■) around 196 ± 14 nm, in the interval 2.3 – 2.5 μm, the average s was 3.6 ± 0.5; coincident with a more formal value for an impurity ions scattering. On the other hand, Sc doped films had s values around 3.2 ± 0.3, with a tendency to diminish for dopant concentration higher than 6 at. %; in this case any dependence with the film thickness was observed; this s value also corresponds to impurity ions scattering. For In doped ZnO crystals, s value of approximately 3 was estimated experimentally [59].

4.2. Photocatalytic Activity of Thin Films Deposited by Aerosol Assisted CVD

Environmental pollution on a global scale has drawn much attention to the necessity of ecological technologies, materials, and processes.

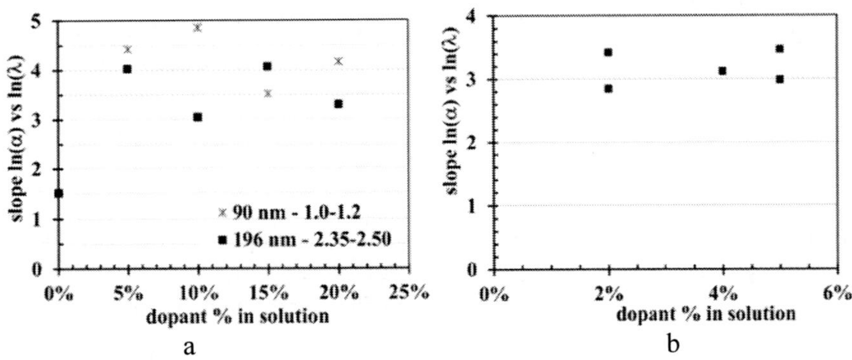

Figure 19. Exponent s of the wavelength dependence of α_{FCA} estimated from the slope of the plot $\ln(\alpha_{FCA})$ vs $\ln(\lambda)$. a) Co doped films: they show a difference in the wavelength interval over which FCA takes place. For thinner (90 ± 11 nm) samples (*), around 1.0 - 1.2 μm, with average s of 4.2 ± 0.6; and for thicker (196 ± 14 nm) samples (■), around 2.3 – 2.5 μm, with average s of 3.6 ± 0.5. b) Sc doped films had s values around 3.2 ± 0.3, with a tendency to diminish for dopant concentration higher than 6 at. %; in this case any dependence with the film thickness was observed.

Since the report of Fujishima [60] in 1972, of the photo-decomposition of water into H_2 and O_2 using an electrode of TiO_2, many works have been done in this area [61]. Early studies were focused on the production of hydrogen from water; afterwards it was reported that UV irradiated semiconductor particles could catalyze varied redox reactions of organic and inorganic substrates. These photocatalytic materials, particularly TiO_2 thin films, have been used to decompose organic contaminants, sterilize surfaces or fluid (gas or liquid) flows [62], deactivation of cancer cells [63], clean-up of soils and oil spills [64], photo-induced superhydrophilicity and the self-cleaning effect [65].

Semiconductors such as TiO_2, ZnO, and Fe_2O_3 are the most used photocatalysts for light-induced redox processes.

The photocatalytic process starts when a photon of sufficient energy excites electrons into the conduction band, leaving a hole in the valence band. The electrons in the conduction band and holes in the valence band can react with electron acceptors and electron donors adsorbed on the semiconductor surface. Prevention or reduction of electron and hole recombination is necessary, to make possible these redox reactions.

Consequently, valence band holes act as powerful oxidants, they can produce hydroxyl radicals ($OH^{\bullet}$); whereas the conduction band electrons are good reductants [66], they can reduce O_2 to superoxide (O_2^-).

These features render the photocatalyst-covered substrate or tubing applicable to environmental protection, self-cleaning, air purification, water purification and sterilization, especially in medical facilities and factories manufacturing pharmaceutical and medical devices.

Recent research is directed to expand the energy interval towards visible light following several approaches, such as:

a)   Metal and nonmetal doping.
b)   Solid solutions.
c)   Composite materials.

Most of photocatalytic studies or applications have been performed with highly efficient powder materials as photocatalyst. However, two main drawbacks exist in this approach:

a)   The turbidity takes place while the photocatalyst in suspension, decreasing the number of photons available for the process.
b)   Final separation steps are necessary to purify the suspension liquid, and recovering or recycling the photocatalyst [67].

Therefore, immobilization of the photocatalyst as a thin film on inert (cheaper) support, such as glass [68], ceramics [69], plastic fiber-optic cable [70] or glass fiber [71], eliminates these drawbacks; nevertheless additional optimization of some characteristics of the photocatalyst are necessary, for example, its surface area.

Moreover, the activation of conventional photocatalyst, such as TiO_2 and ZnO, take place only under UV illumination [72]. TiO_2 and ZnO absorb only less than 10 % of sunlight, limiting its applicability with solar or ordinary lighting systems.

Significant progress has been made to render oxide-based materials as a visible light photocatalyst. In order to absorb visible light, the band gap (BG) of oxide photocatalyst has to be narrowed, which can be achieved by doping with transition metal ions, e. g., Cu, Ni, Co, Cd, Ag, Mn [73].

Another alternative is the use of a multilayered structure with several sub-gaps covering different regions of the solar spectrum. At the same time, recent studies using heterostructures, where semiconductors with different carrier band structures are superimposed, show the increase of the charge carriers' life time and thus improving the photo activity [72].

In this section the photocatalytic properties of Zinc oxide thin films are presented; it was selected due to its advantageous characteristics. ZnO is a wide band gap semiconductor with the capability to degrade a wide range of pollutants with high photocatalytic efficiency [74, 75]. ZnO has been considered as a promising photocatalyst candidate; in addition it exhibits high mechanical stability, non-toxicity along with its abundance in nature, which makes it a lower cost material [73].

The photocatalytic activity of oxide films can be tested by the discoloration of organic dye (methylene blue or methyl orange) in aqueous solution. The initial concentration of solution probe was fixed at 10^{-5} $mol.dm^{-3}$. The film arrangements were introduced inside of an irradiation chamber and exposed to a UV-A lamp (UV BL-15) at 365 nm with time intervals of 120 and 240 minutes. Absorbance spectra of irradiated solution in contact with film were measured in a Perkin Elmer Lambda 35 or Cary 5000 UV-VIS-NIR spectrophotometers; where changes of characteristic absorbance peak were monitored as indicators of photocatalytic efficiency.

4.2.1. Photocatalytic Activity of Internally Coated Tubular Reactor with ZnO Films Deposited by Aerosol Assisted CVD

Thin films of photocatalytic ZnO were deposited on the internal surface of fused silica tubing by a reproducible AACVD technique. The experimental set up and details of the synthesis were reported previously [16].

Several ZnO film coated tubings of 7 cm of length were used as batch reactors for photocatalytic activity measurements. The photocatalytic activity of ZnO films were tested by the discoloration of analytic grade methyl orange (MO) in aqueous solution.

The initial concentration of MO solution was round 1.0×10^{-5} mol.dm^{-3}. The tubing-film arrangements were introduced inside of an irradiation chamber and exposed to a UV-A lamp (UVL-36) at 365 nm at different time intervals.

Changes of MO absorbance peak at 664 nm were monitored to correlate with photocatalytic efficiency of the films.

On the other hand, it is well stablished theoretically, that the photocatalytic reaction rate on the surface of the film is proportional to the surface carrier concentrations [76]. The influence of the film thickness and other material dependent parameters on the photcatalytic activity were evaluated; for this task theoretical calculations of photogenerated charge carrier distribution inside the irradiated photocatalyst and surface carrier concentration were performed, considering external irradiation of the tubing.

Due to the very small thickness of the film ($\sim$ 100 nm) compared to the tubing radii ($\sim$ 4 mm), a one dimensional model was formulated [77] as a good approximation of the cylindrical geometry.

The model considers the solution of the one dimensional stationary continuity equation, taking the origin of spatial coordinate ($x = 0$) in the film-tubing interface and $x = d$ at the film surface:

$$D\frac{d^2 n(x)}{dx^2} - \frac{n(x)}{\tau} + G(x) = 0 \tag{18}$$

where, $n(x)$ is the carrier concentration (either electrons or holes) at a distance x from the film-tubing interface; D is the diffusion coefficient of the charge carriers; τ is their lifetime; and $G(x)$ is carrier photogeneration rate per unit volume at x.

The carrier photogeneration function $G(x)$ can be expressed as [77]:

$$G(x) = \chi \cdot \alpha(\lambda) \cdot F_{ph} \cdot e^{-\alpha(\lambda)\cdot x} \tag{19}$$

where, χ is the quantum efficiency to create an electron–hole (e–h) pair per absorbed photon; $\alpha(\lambda)$ is the absorption coefficient of the ZnO film, at the wavelength λ; and F_{ph} is the incident photon flux of wavelength λ (photons cm^{-2}s^{-1}).

Solution of the continuity equation can be found elsewhere [77]. Utilized parameters were: $\chi \sim 1$; $\alpha(365$ nm$) \sim 2.0 \times 10^5$ cm^{-1}; $\tau \sim 1.4 \times 10^{-8}$ s [78]; D ~ 2 - 5 cm^{2}s^{-1}; and $F_{ph} \sim 1.8 \times 10^{15}$ cm^{-2} s^{-1}.

Diffusion coefficient was estimated according to the Einstein relation:

$$D = \frac{kT}{q_e}\mu \tag{20}$$

where k is the Boltzmann constant, T is the temperature, q_e is the fundamental charge, and μ is the mobility of the carriers; we have considered ambient temperature (T ~ 300 K) and mobilities for n-type ZnO on the order of 100 – 200 $cm^2.V^{-1}.s^{-1}$. Another parameter was the coefficient of surface recombination s ~ 1.0 x 10^3 $cm.s^{-1}$ [79].

With these parameters, the diffusion length (L = $\tau^{1/2}.D^{1/2}$) of the photocarriers was of the order of 1.3 - 2.7 μm.

Figure 20 presents the calculated surface carrier density as a function of a) film thickness, compared to relative activity of several ZnO thin films of different thickness; b) recombination time τ, c) diffusion coefficient D, and d) absorption coefficient α.

An optimum film thickness around 80 - 90 nm was obtained, for which the surface carrier density is maxima, (see Figure 20a); consequently the photocatalytic activity should be also maxima. This behavior is a consequence of the external irradiation of the tubing-film system because the surface carrier is determined by:

a)   The rate of photo-generated in the film, which is higher at the interface film-tubing (x = 0).
b)   the rate of carrier diffusion flow to the surface (x = d).
c)   by their decay through surface recombination and reaction pathways.

Qualitatively surface carrier density matches with experimental relative photocatalytic activity reported elsewhere [77] (see Figure 20a).

The influence of critical parameters, such as τ, D·and α, on the surface carrier density can be evaluated following this model. These quantities depend on the electronic structure of the materials; they determine photogeneration and the diffusion length of the photocarriers, and consequently the surface carrier density and the photocatalytic activity.

Figures 20 b), c) and d) show the calculated surface carrier density n(d), for a film of 70 nm of thickness, as a function of recombination time τ, diffusion coefficient D of charge carriers and absorption coefficient of the film α, respectively. Also shown for a fixed film thickness that the surface carrier density increases with the increase of these parameters up to a certain limit, and then it stays almost constant, because of the limited carrier generation in the fixed volume of the film. In this high diffusion limit the carrier concentration is approximately constant in all the volume of the film.

However, the variation of these material dependent parameters (τ, D and α) has an important influence in the optimum film thickness, for which surface carrier density is maxima. Table 13 presents calculated optimum thickness for several values of τ, D and α.

It is also shown that absorption coefficient and recombination time have a strong influence in the surface carrier density and optimum thickness; in contrast and rather surprisingly, diffusion coefficient has little influence in both quantities (see Table 13).

The photonic efficiency [76] (number of discolored molecules per incident photon) of ZnO coated tubing was evaluated to assess the influence of experimental conditions: a) MO concentration; b) irradiation time; c) irradiation intensity. For solution concentration dependency the heterogeneous character of the system was confirmed by a typical Langmuir-Hinshelwood kinetics. It was found out that for diluted solutions ($C < 0.1$ mmol L^{-1}) the reaction is first order, whereas for higher concentrations ($C > 0.2$ mmol L^{-1}), the reaction rate is maximum with zero order.

Photonic efficiency decreased with the irradiation time, consistent with the decrease of solution concentration as the irradiation time increased.

Table 13. Influence of recombination time τ, diffusion coefficient D, and absorption coefficient α in the surface carrier density n(d) and optimum film thickness d_{opt}. Considering an external irradiated film-tubing system

τ [ns]	D [m^2.s^{-1}] 10^{-4}	α [m^{-1}] 10^7	n(d) [cm^{-3}] 10^{12}	d_{opt} [nm]
1.4	3.9	2	0.96	30
14	3.9	2	3.41	80 - 90
140	3.9	2	5.78	170 - 180
14	0.39	2	3.40	80 - 90
14	39	2	3.41	80 - 90
14	3.9	0.2	0.96	350
14	3.9	20	5.76	20

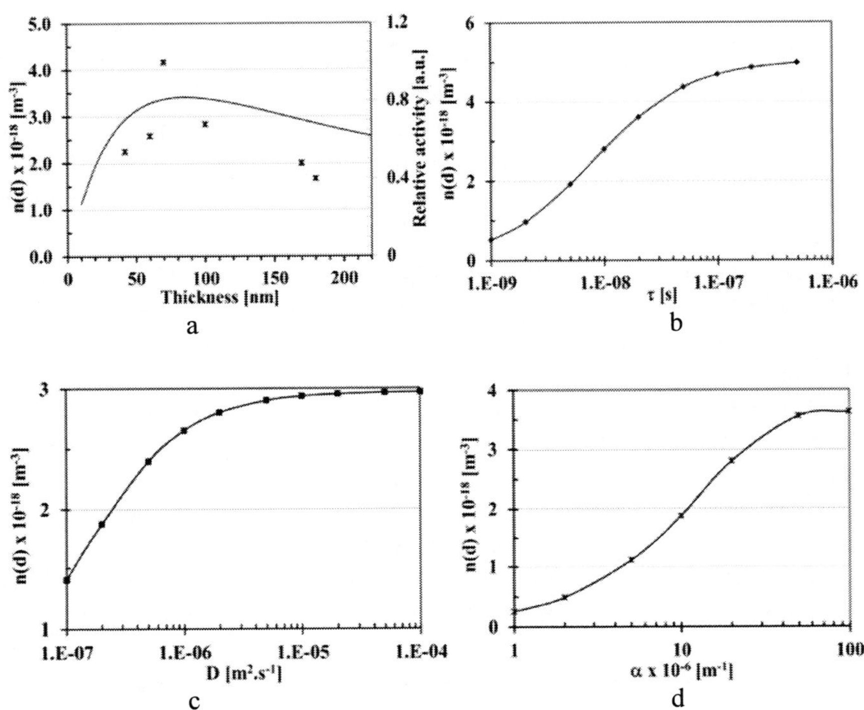

Figure 20. Calculated surface carrier density as a function of a) film thickness, included is the relative photocatalytic activity of several ZnO thin films of different thickness, for comparison; b) τ recombination time; and c) D diffusion coefficient and d) α absorption coefficient.

Finally, photonic efficiency decreases initially with the irradiation intensity, then it remained almost constant; this behavior can be attributed to the increase of the rate of recombination of electron-hole pairs as the incident intensity increased.

4.2.2. Photocatalytic Activity of ZnO Based Thin Films Deposited on Flat Substrates by Aerosol Assisted CVD

Thin films of photocatalytic ZnO were deposited onto borosilicate glass substrates by a reproducible AACVD technique, in some samples a buffer film of ZnO was previously deposited as a diffusion barrier. The experimental set up and details of the synthesis were reported previously [18, 80]. Several ZnO based films were tested by the decoloration of methylene blue in aqueous solution. The initial concentration of MB solution was fixed at 10^{-5} mol.dm^{-3}.

In order to limit the contact area of the MB solution with the films, a Teflon ring of 2.5 cm of internal diameter and 1.1 cm^3 of capacity was firmly placed on the surface of the samples.

The first group correspond to Co doped ZnO films. Figure 21 shows the absorbance of MB solutions after 4 h of irradiation with the ZnO based films. An optimum Co concentration of 5 at. % for maxima discoloration, i.e., photocatalytic activity, was obtained (see Figure 21a). Figure 21b presents the absorbance of MB solutions irradiated on 5 at. % Co doped films of different thickness.

It was discovered that photocatalytic activity increases as the film thickness increases up to around 220 nm. This fact is contrary to the case of an external irradiated film-tubing system, discussed in the last unit, where an optimum thickness of around 70 nm was found.

The second type of ZnO based films corresponds to a bilayered ZnO / ZnO-CuOx film structure, including a buffer layer of ZnO as a diffusion barrier of some noxious elements from the borosilicate substrate, mainly Na. In the second layer the Cu/(Zn+Cu) atomic ratio varied from 0.1 to 0.9. Zincite and Tenorite phases were found for the respective Zn and Cu oxides in the composite films. No significant degradation of MB was observed neither for films of low or high Cu concentrations. However, those samples with Cu concentrations around 50 at. %, a considerable MB discoloration (82 to 93 %) was reached, similar to that obtained for undoped ZnO sample.

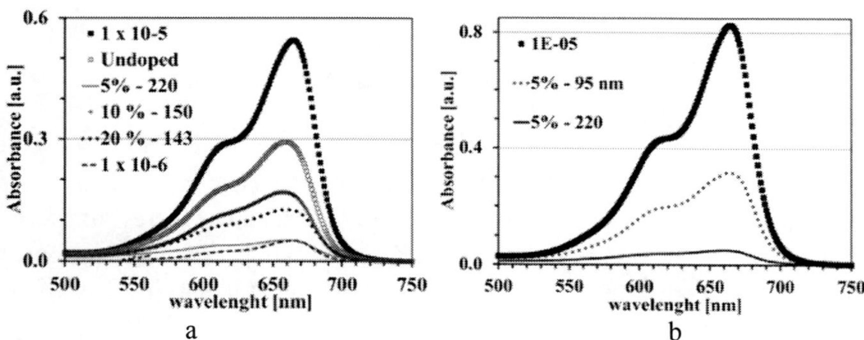

Figure 21. Absorbance measurement of irradiated MB solution in contact with photocatalytic Co doped ZnO films, compared to the absorbance of reference solutions of 10^{-5} and 10^{-6} mol.dm^{-3}. a) as a function of Co concentration. b) as a function of thickness for Co doped films of 5 at. %.

Table 14. Cu content in solution, total thickness, absorbance at 350 nm and photocatalytic activity of bilayered Zn-Cu oxide thin films

Cu/(Zn+Cu) at. %	Total thickness [nm]	Absorbance at ~365 nm [%]	Photocatalytic activity [%]
Undoped	86	53	95
40	52	32	93
50	51	52	82
60	45	41	93

Slightly higher activity of undoped sample compared to bilayered Zn-Cu oxide samples can be explained due to the higher thickness and higher absorbance, at the wavelength of irradiation (UV-A light ~ 365 nm), of undoped film.

Lower absorbance of Zn-Cu oxide films is due to their high reflectance in the near UV and visible interval. Consequently, the lower absorbance indicates that the quantum efficiency (number of discolored molecules per absorbed photon) of Zn-Cu oxide films is higher than that of the undoped ZnO film.

Table 14 presents the main characteristics of the films with highest photocatalytic activity as deduced from the discoloration of a MB aqueous solution. In this case MB discoloration test was performed for 4 h of irradiation.

4.3. As Adsorption on Aerosol Assisted Synthesis of Nanostructured Materials

Arsenic is one of the most toxic and carcinogenic chemical elements when consumed in quantities greater than 10 ppb. Recently, there has been a growing interest towards the use of iron based nanoparticles for adsorptive removal of a variety of organic and inorganic contaminants [81], including organic dyes or heavy metal cations from water, including As [82].

Nanomaterials have several advantages over other materials, such as activated [83] or mesoporous alumina [84], mainly due to their higher adsorption capacity, generation of lower quantities of contaminated sludge, and less time for reaching the adsorption equilibrium [85]. In addition, adding magnetic characteristics to these nanomaterials, highly efficient magnetic separation and recovery can be possible.

Arsenic removal tests were performed using hollow spherical magnetite (HSM) nanoparticles [22], on different aqueous solutions of As^{+3} at 0.08 ppm and As^{+5} at 0.05 ppm. For both As^{+3} and As^{+5} solutions, 10 mg of magnetite nanoparticles were added to 100 mL of tri-distilled water. Nanoparticles were left in contact for different periods of time (1, 3, 5, 10, 20 and 30 minutes) under constant stirring.

Afterwards, nanoparticles were collected with magnets; and the decanted liquid was analysed by atomic absorption spectrometry to measure both, arsenic and iron concentrations. The experiments were repeated to verify the reproducibility of the results.

Results were remarkable since removal of almost 100 % of both, As^{+3} and As^{+5} ions, in short contact time (~ 1 min) were obtained. Thus, arsenic removal carried out very fast, which means that the affinity of arsenic by iron is very strong. Besides this affinity, the rapid adsorption of the metal ions on the HSM nanoparticles is also caused by the high specific area of the adsorbents. In the literature there are other reports where arsenic is removed by iron based nanoparticles [81, 82, 86, 87], with efficiencies in the range of 70 – 99.9 %, but with exposition times of several minutes up to 120 minutes. Then, the results presented in this work are certainly notable.

Furthermore, another advantage of using this type of particles in the removal of arsenic is the magnetic capability of nanoparticle recollection from the tested solution; due to the paramagnetic nature of the particles, they are easily separated from the tested solution with the help of magnets. This was confirmed since the iron content in all tested solutions was found to be undetectable. Therefore, this feature makes more attractive the use of this type of nanoparticles for the removal of heavy metal ions, since the sludge produced is easily separated from the treated water.

CONCLUSION

Nanostructured oxides as thin films and nanoparticles were obtained by aerosol assisted chemical vapor deposition technique. Thin films of several undoped or doped oxides, mainly consisting of ZrO_2, ZnO and TiO_2, were deposited as single layer, multi-layered or composite materials. Deposited films were polycrystalline of high quality, uniform, non-light scattering and well adhered to the substrate. On the other hand, hollow spherical magnetite nanoparticles were also obtained.

A detailed microstructural characterization of these materials was performed by electron microscopy and x-ray diffraction.

Microstructural characteristics of the materials such as composition, crystalline phases, grain morphology, crystallite size, lattice parameters were obtained as a function of several deposition parameters.

In the case of yttria stabilized ZrO_2 films, a polycrystalline multi-layered microstructure of alternated dense and porous agglomerated crystallites were observed. Small crystallite size, around 7 nm, was observed by transmission electron microscopy in accordance to that calculated by peak broadening in GIXRD. The multi-layered microstructure was generated as a consequence of the periodic nozzle sweep along the substrate.

For ZnO based films and multilayers deposited by aerosol assisted CVD, dopant to zinc ratio in the film was dependent of the particular dopant precursor employed; for example, Co/Zn at. ratio in film was larger than that in solution, in contrast for Sc, Mg and Cu this ratio was lower. Co and Mg dopant showed very high solubility (> 30 at. %) in the ZnO structure, in contrast with other dopants, whose solubility limit was at only a few at. %. Cross sectional microstructure analysis showed that well crystallized close packed grains composed the films; crystallite size was estimated around 30 ± 10 nm, independently of the type of dopant.

For nanostructured magnetite synthesized by aerosol assisted CVD, spherical hollow nanoparticles were obtained; with overall characteristics slightly dependent on the synthesis temperature, as follows: external diameter 300 - 380 nm, crust thickness 26 - 27 nm, crystallite size 19 - 25 nm, porosity 30 - 59 % and specific surface area $66 - 47$ $m^2.g^{-1}$. The porous microstructure of the hollow nanoparticles causes the component crystallites to perform as isolated spherical nanoparticles in regard to surface related phenomenon, for example photocatalysis or adsorption removal of pollutants. Then, hollow spherical nanoparticles (~300 nm of diameter) of this work have advantageous characteristics concerning particle manipulation or recollection and, on the other hand, the component (agglomerated) magnetite crystallites (~20 nm of size) perform as isolated spherical nanoparticles with regard to their specific surface area.

Optical characterization of the yttria stabilized zirconia thin films permitted the determination of optical band gap energy E_g in the interval of 5.9 – 6.0 eV, depending of the Y concentration. High values of band gap can be attributed to quantum confinement, due to the small crystallite size (~7 nm). Dispersion energy E_d of the single oscillator model of refractive index n was evaluated from its spectral dependence, values around 24 ± 1 eV were obtained;

with this result, a covalent character of the interatomic bond ($\beta \sim 0.37$) was deduced considering 4 effective valence electrons per anion.

Optical properties of ZnO based films were also evaluated. Absorption coefficient of Co doped films show: a) fundamental inter-band absorption at the band gap energy $E_g \sim 3.3$ eV; b) Co^{+2} ions associated bands at ~ 2 eV and ~ 1 eV; and c) free carrier absorption in the NIR generated by intra-band transitions. Fundamental inter-band absorption for Co and Mg doped films showed a linear correlation of E_g vs dopant x concentration in the film. The following correlations were found: $E_g^{Co}(x) = 3.30 + 0.5x$ and $E_g^{Mg}(x) = 3.28 + 1.6x$ $[eV]$. On the other hand, the existence of the Co^{+2} ions associated bands indicate that Co dopant is actually substituting Zn atoms in the wurtzite structure. However, an important increase of Urbach width was also observed for Co doped films. Urbach absorption tail is generated by heavy doping of degenerate semiconductors, attributed to strong internal fields arising from ionized dopants or defects. Finally, free carrier absorption is generated by photons of energy bellow E_g. The spectral dependence of free carrier absorption with wavelength indicates that the predominant scattering mechanism was due to impurity ions.

Photocatalytic activity of selected ZnO based thin films was evaluated by discoloration of methylene blue solutions. For ZnO internally coated tubular reactor a photo-generated carrier distribution model was formulated solving the one-dimensional continuity equation. With this model the influence of film thickness and other parameters dependent on the electronic structure of the photocatalyst, over the photocatalytic activity was evaluated, considering external irradiation of the reactor. An optimum thickness of around 80 nm was deduced to get maximum surface carrier density, using similar conditions as those of the experimental tests. Following the model, recombination time and absorption coefficient were the most influential parameters on the surface carrier density. Photonic efficiencies for films deposited onto flat borosilicate substrates, Co doped ZnO and a bilayered ZnO/CuO-ZnO films were also tested.

In the case of Co doped films an optimum Co/(Zn+Co) at. ratio in solution of 5 % was obtained, for a ~ 90 % of discoloration of a 10^{-5} mol.dm^{-3} MB solution. For bylayered CuO-ZnO film, an optimum at. ratio Cu/(Zn+Cu) of 50 % was found, obtaining around of ~ 90 % of discoloration of a 10^{-5} mol.dm^{-3} MB solution.

On the other hand, hollow spherical magnetite nanoparticles were used for the removal of As from aqueous solutions. Tests were performed using of As^{+3} at 0.08 ppm and As^{+5} at 0.05 ppm, with 10 mg of magnetite nanoparticles in

100 mL of tri-distilled water. Remarkable results were obtained; removal of almost 100 % of both, As^{+3} and As^{+5} ions, in short contact time (~ 1 min).

ACKNOWLEDGMENTS

The authors would like to thank A. Sáenz-Treviso, B. Monárrez-Cordero, P.G. Hernández-Salcedo, E. Torres-Moye, K. Campos-Venegas, R. Ochoa, S. Miranda, L. de la Torre-Sáenz, D. Lardizabal, M. Lugo-Ruelas, O. Esquivel-Pereyra, and G. Rivas-Amézaga for experimental assistance. This work was partially supported by project CONACYT-SEP 2008-106655.

REFERENCES

[1]   Granqvist, C. G. *Sol. Ener. Mat. Sol. C.* 2007, 91, 1529-1598.

[2]   Coey, J. M. D.; Venkatesan, M.; Fitzgerald, C. B. *Nature Materials*, 2005, 4, 173179.

[3]   Prellier, W.; Fouchet, A.; Mercey, B. *Phys. Condens. Matter.* 2003, 15, R1583–R1601.

[4]   Dietl, T.; Ohno, H.; Matsukura, F.; Cibert, J.; Ferrand, D. *Science.* 2000, 287, 1019–1022.

[5]   Matsumoto, Y.; Murakami, M.; Shono, T.; Hasegawa, T.; Fukumura, T.; Kawasaki, M.; Ahmet, P.; Chikyow, T.; Koshihara, S.; Koinuma, H. *Science.* 2001, 291, 854–856.; Ueda, K.; Tabata, H.; Kawai, T. *Appl. Phys. Lett.* 2001, 79, 988–990.; Ogale, S. B.; Choudhary, R. J.; Buban, J. P.; Lofland, S. E.; Shinde, S. R.; Kale, S. N.; Kulkarni, V. N.; Higgins, J.; Lanci, C.; Simpson, J. R.; Browning, N. D.; Das Sarma, S.; Drew, H. D.; Greene, R. L.; Venkatesan, T. *Phys. Rev. Lett.* 2003, 91, 077205.

[6]   Novoselov, K. S.; Geim, A. K.; Morozov, S. V.; Jiang, D.; Zhang, Y.; Dubonos, S. V.; Grigorieva, I. V.; Firsov, A. A. *Science.* 2004, 306, 666-669.

[7]   Regis, Y. N.; Konstantinos Spyrou, G.; Rudolf, P. *J. Phys. D. Appl. Phys.* 2010, 43, 374015 (19 pp.). Reina, A.; Jia, X.; Nezich, D.; Son, H.; Bulovic, V.; Dresselhaus, M. S.; Kong, J. *Nano Lett.* 2009, 9(1), 30-35.

[8]   Vyas, J. D.; Choy, K. L. *Mater. Sci. Eng.* 2000, A277, 206–212.

[9]   Evans, A.; Bieberle-Hütter, A.; Rupp, J. L. M.; Gauckler, L. J. *J. Power Sources.* 2009, 194, 119–129.

[10] Korotcenkov, G. *Sensors and Actuat.* B. 2005, 107, 209–232.

[11] Seshan, K. *Handbook of thin-film deposition processes and techniques*; 2nd Edition; Noyes Publications: Norwich, NY, 2002; Ch. 1.

[12] Peter, E.; Martin, M. *Handbook of deposition technologies*, 3rd Edition, Elsevier: Oxford, UK, 2010; Ch. 1.

[13] McNatt, J.; Dickman, J.; Jin, M.; Banger, K.; Kelly, C.; González, A.; Rockett, A. *NASA/TM.* 2006-214445.

[14] Amézaga-Madrid, P.; Antúnez-Flores, W.; Monárrez, I.; González-Hernández, J.; Martínez-Sánchez, R.; Miki-Yoshida, M. *Thin Solid Films.* 2008, 516, 8282-8288.

[15] Ramos-Cano, J.; Hurtado-Macías, A.; Antúnez-Flores, W.; Fuentes-Cobas, L.; González-Hernández, J.; Amézaga-Madrid, P.; Miki-Yoshida, M. *Thin Solid Films.* 2013, 531, 179-184.

[16] Miki-Yoshida, M.; Collins-Martínez, V.; Amézaga-Madrid, P.; Aguilar-Elguezabal, A. *Thin Solid Films.* 2002, 419, 60-64.

[17] Amézaga-Madrid, P.; Hurtado-Macías, A.; Antúnez-Flores, W.; Estrada, F.; Pizá-Ruiz, P.; Miki-Yoshida, M. *J. Alloy. Compd.* 2012, 536, S412–S417.

[18] Amézaga-Madrid, P.; Antúnez-Flores, W.; Sáenz-Hernández, R. J.; Martínez-Sánchez, R.; Miki-Yoshida, M. *J. Alloy. Compd.* 2009, 483, 410-413.

[19] Amézaga-Madrid, P.; Antúnez-Flores, W.; Ledezma-Sillas, J. E.; Murillo-Ramírez, J. G.; Solís-Canto, O.; Vega-Becerra, O. E.; Martínez-Sánchez, R.; Miki-Yoshida, M. *J. Alloy. Compd.* 2011, 509, S490–S495.

[20] Amézaga-Madrid, P.; Nevárez-Moorillon, G.; Orrantia-Borunda, E.; Miki-Yoshida, M. *FEMS Microbiol. Lett.* 2002, 211, 183-188.

[21] Miki-Yoshida, M.; Andrade, E. *Thin Solid Films.* 1993, 224, 87-96.

[22] Monárrez-Cordero, B.; Amézaga-Madrid, P.; Antúnez-Flores, W.; Leyva-Porras, C.; Pizá-Ruiz, P.; Miki-Yoshida, M. *J. Alloys Compd.* 2013, 586, S520–S525.

[23] Esparza-Ponce, H. E.; Reyes, A.; Antúnez-Flores, W.; Miki-Yoshida, M. *Mater. Sci. Eng.* A 2003, 343, 82-88.

[24] Vick, D.; Brett, M. J. *J. Vac. Sci. Technol.* A. 2006, 24, 156–164.; Malac, M.; Egerton, R. F. *J. Vac. Sci. Technol.* A. 2001, 19, 158–166.; Lakhtakia, A.; R. Messier, R. *Sculptured Thin Films: Nanoengin. Morph. Optics*, SPIE Press, Bellingham, US, 2005.

[25] Choy, K. L. *Prog. Mater. Sci.* 2003, 48, 57–170.

[26] G. Blandenet, M. Court, Y. Lagarde, *Thin Solid Films* 1981, 77, 81.

[27] Amézaga-Madrid, P.; Antúnez-Flores, W.; González-Hernández, J.; Sáenz-Hernández, J.; Campos-Venegas, K.; Solís-Canto, O.; Ornelas, C.; Vega-Becerra, O.; Martínez-Sánchez, R.; Miki-Yoshida, M. *J. Alloys Compd.* 2010, 495, 629-633.

[28] Wu, S. G.; Zhang, H. Y.; Tian, G. L.; Xia, Z. L.; Shao, J. D.; Fan, Z. X. *Appl. Surf.* 2006, 253, 1561–1565.

[29] Joint Committee on Powder Diffraction Standards, Powder Diffraction File, International Center for Diffraction Data, Swarthmore, PA, 2011, card 03-065-0461.

[30] Joint Committee on Powder Diffraction Standards, Powder Diffraction File, International Center for Diffraction Data, Swarthmore, PA, 2011, card 00-030-1468.

[31] Cullity, B. D.; Stock, S. R. *Elements of X-Ray Diffraction*, 3rd ed., Prentice Hall, NJ, 2001, pp. 170.

[32] Ducros, C.; Cayron, C.; F. Sanchette. *Surf. Coat Technol.* 2006, 201, 136–142.

[33] Paraguay-Delgado, F.; Morales, J.; Estrada, W.; Andrade, E.; Miki-Yoshida, M. *Thin Solid Films.* 2000, 366, 16-27.

[34] Gómez, H.; Maldonado, A.; Asomoza, R.; Zironi, E. P.; Cañeta, S.; Palacios, J. *Thin Solid Films.* 1997, 293 117-123.

[35] Schaedler, T. A.; Gandhi, A. S.; Saito, M.; Rühle, M.; Gambino, R.; *J. Mater. Res.* 2006, 21, 791-799.

[36] Janotti, A.; Van de Walle, C. *Rep. Prog. Phys.* 2009, 72, 126501 (29 pp.).

[37] Joint Committee on Powder Diffraction Standards, Powder Diffraction File, International Center for Diffraction Data, Swarthmore, PA, 2011, card 01-079-0205.

[38] Yoo, Y. Z.; Jin, Z. W.; Chikyow, T.; Fukumura, M.; Kawasaki; Koinuma, H. *J. Appl. Phys.* 2001, 90, 4246-4250.

[39] Yano, M.; Koike, K.; Sasa, S.; Inoue, M. ZnO/ZnMgO Heterojunction FETs, in Zinc Oxide Bulk, *Thin Films and Nanostructures Processing, Properties, and Applications*, Elsevier 2006; Ch. 11, 371-414.

[40] Xiong, Y.; Kodas, T. *J. Aerosol Sci.* 1993, 24, 893-908.

[41] Lenggoro, I. W.; Hata, T.; Iskandar, F.; Lunden, M. M.; Okuyama, K. *J. Mater. Res.* 2000, 15(3), 733-743.

[42] Joint Committee on Powder Diffraction Standards, Powder Diffraction File, *International Center for Diffraction Data*, Swarthmore, PA, 2011, card 01-089-0691.

[43] Palik, E. D., *Handbook of Optical Constants of Solids*, 1st ed., Academic Press Inc., CA, 1985, pp. 21.

[44] Stenzel, O., *The Physics of Thin Films Optical Spectra*, Springer-Verlag, Berlin, 2005, 42–45.

[45] Kireev, P. S. *Semiconductor Physics*, MIR, Moscow, 1966, Ch. 8, 555.

[46] Sharma, R.; Sehrawat, K.; Wakahara, A.; Mehra, R. M. *Appl. Surf. Sci.* 2009, 255, 5781–5788.

[47] Diaz-Parralejo, A.; Caruso, R.; Ortiz, A. L.; Guiberteau, F. *Thin Solid Films.* 2004, 458, 92–97.

[48] Marques, A. C.; Chevalier, P. *Handbook of Nanostructured Thin Films and Coatings, Functional Properties.* Ed. Sam Zhang. CRC Press. 2010, Ch. 6, 255.

[49] Heiroth, S.; Ghisleni, R.; Lippert, T.; Michler, J.; Wokaun, A. *Acta Materialia.* 2011, 59, 2330–2340.

[50] Wemple, S. H.; DiDomenico, M. Jr. *Phys. Rev. B.* 1971, 3, 1338.

[51] Zhao, S.; Ma, F.; Xu, K. W.; Liang, H. F. *J. Alloys Compd.* 2008, 453, 453–457.

[52] Hidekazu, I.; Krause, M.; Hoche, T.; Patzig, C.; Hu, Y. *J. Phys. Condens. Matter.* 2013, 25 165505-165513.

[53] Kasam, S.; Capper, P. *Springer Handbook of Electronic and Photonic Materials.* Part A/3. Springer 2006, 47-77.

[54] Hirohide Nakamatsu, Takeshi Mukoyama, Hirohiko Adachi, *Chemical Physics Letters* 247 (1995) 168-172.

[55] Ramana, C. V.; Vemuri, R. S.; Fernandez, I.; Campbell, A. L. *Appl. Phys. Lett.* 2009, 95,2330-2340.

[56] Camagni, P.; Galinetto, P.; Samoggia, G.; Zema, N. *Sol. State Commun.*, 1992, 83, 943-947.

[57] Kim, K. J.; Park, Y. R. *Appl. Phys. Lett.* 2002, 81, 1420 – 1422.

[58] Gorschlüter, A.; Merz, H. *Phys. Rev. B.* 1194, 49, 17 293-17302.

[59] Weiher, R. L. *Phys. Rev.* 1966, 152, 736 –739.

[60] Fujishima, A.: Honda, K. *Nature.* 1972. 238, 37–38.

[61] Fujishima, A.; Zhang, X.; Tryk, D. A. *Inter. J. Hydrog. Energ.* 2007, 32, 2664-2672.

[62] Ireland, J. C.; Klostermann, P.; Rice, E. W.; Clark, R. M. *Appl. Environ. Microbiol.* 1993, 591668–1670.

[63] Cai, R. X.; Kubota, Y.; Shuin, T.; Sakai, H.; Hashimoto, K.; Fujishima, A. *Cancer Res.* 1992, 52, 2346–2348.

[64] Nair, M.; Luo, Z. H.; Heller, A. *Ind. Eng. Chem. Res.* 1993, 32, 2318–2323.

[65] Anpo, M.; Takeuchi, M.; Yamashita, H.; Hirao, T.; Itoh, N.; Iwamoto, N. *Proc. 4th Int. Conf. ECOMATERIAL Gifu.* Gifu City, Japan: Sci. Technol. Agency. 1999, 333–36.

[66] Gratzel, M. *Heterogeneous Photochemical Electron Transfer.* Boca Raton, FL: CRC Press. 1989, 159.

[67] Saragiotto, L. M.; Alves, H. J., Andreo, O. A.; Macedo, C. M. *Dyes Pigm.* 2008, 76, 525-529.

[68] Krýsa, J.; Novotná, P.; Kment, S.; Mills, A. *J. Photochem. Photobiol.* A. 2011, 222, 81-86.

[69] Zhang, P.; Tian, J.; Xu, R.; Ma, G. *Appl. Surf. Sci.* 2013, 266141-147.

[70] Kulkarni, A.; Lee, J.; Nam, J.; Kim, T. *Sens. Actuat.* B. 2010, 150 154-159.

[71] Peilland, N. J.; Hoffmann, M. R. *Environ. Sci. Technol.* 1996, 30, 2806-2812.

[72] Sapkal, R. T.; Shinde, S. S.; Waghmode, T. R.; Govindwar, S. P.; Rajpure, K. Y.; Bhosale, C. H. *J. Photochem. Photobiol. B: Biol.* 2012, 110 15–21.

[73] Anandan, S.; Ohashi, N.; Miyauchi, M. *Appl. Catal. B: Environ.* 2010, 100, 502–509.

[74] Anpo, M.; Dohshi, S.; Kitano, M.; Hu, Y.; Takeuchi, T.; Matsuoka, M. *Annu. Rev. Mater. Res.* 2005, 35, 1–27.

[75] Amornpitoksuk, P.; Jongnavakit, J.; Suwanboon, S.; Ndiege, N. *Appl. Surf. Sci.* 2012, 258 8192–8198.

[76] Emeline, A.; Salinaro, A.; Serpone, N. *J. Phys. Chem. B.* 2000, 104, 11202-11210.

[77] Ríos-Valdovinos, E.; Amézaga-Madrid, P.; Antúnez-Flores, W.; Pola-Albores, F.; Pizá-Ruiz, P.; Miki-Yoshida, M. *Mater. Sci. Engin. B.* 2014, 179, 41– 47.

[78] Özgür, Ü.; Morkoç, Optical properties of ZnO and related alloys, in Zinc Oxide Bulk, *Thin Films and Nanostructures Processing, Properties, and Applications*, Elsevier. 2006, Ch. 5, 228.

[79] Lopatiuk, O.; Chernyak, L.; Osinsky, A.; Xie, J. Q.; Chow, P. P. *Appl. Phys. Lett.* 2005, 87, 162103–162105.

[80] Sáenz-Trevizo, A.; Amézaga-Madrid, P.; Pizá-Ruiz, P.; Solís-Canto, O.; Ornelas-Gutiérrez, C.; Pérez-García, S.; Miki-Yoshida, M. in press *J. Alloys Compds.* 2014.

[81] Cheng, Z.; Van Geen, A.; Louis, R.; Nikolaidis, N.; Bailey, R.; *Environ. Sci. Technol.*, 2005, 39, 7662–7666.

[82] Chandra, V.; Park, J.; Chun, Y.; Lee, J. W.; Hwang, I.Ch.; and Kim, K. S.; *ACS Nano*, 2010, 4, 3979–3986.

[83] Lin, T.-F. and Wu, J.-K., *Water Research*, 2002, 35, 2049-2057.

[84] Kim, Y.; Kim, C.; Choi, I.; Rengaraj, S.; and Yi, J.; *Environ. Sci. Technol.* 2004, 38, 924-931.

[85] Shen, Y. F.; Tang, J.; Nie, Z. H.; Wang, Y. D.; Ren, Y.; Zuo, L.; *Separation and Purification Technology* 2009, 68, 312–319.

[86] Tuutijärvi, T.; J. Lu, J.; Sillanpää, M.; Chen, G.; *Journal of Hazardous Materials* 2009, 166, 1415–1420.

[87] Zouboulis, A. and Katsoyiannis, I., *Ind. Eng. Chem. Res.*, 2002, 41, 6149–6155.

INDEX

D